101 Brilliant Things For Kids To Do With Science

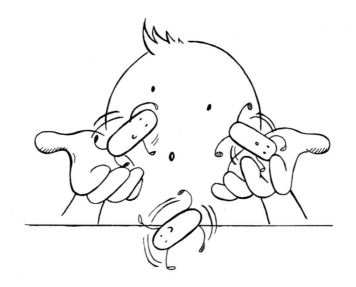

101 Brilliant Things For Kids To Do With Science

DAWN ISAAC

PHOTOGRAPHY BY KATE WHITAKER

Kyle Books

An Hachette UK Company
www.hachette.co.uk

First published in Great Britain in 2017 by
Kyle Books, an imprint of Kyle Cathie Ltd
Carmelite House
50 Victoria Embankment
London EC4Y 0DZ
www.kylebooks.co.uk

This edition published in 2020

ISBN: 9780857838964

Distributed in the US by Hachette Book Group, 1290
Avenue of the Americas, 4th and 5th Floors, New York, NY
10104

Distributed in Canada by Canadian Manda Group, 664
Annette St., Toronto, Ontario, Canada M6S 2C8

Project Editor: Tara O'Sullivan
Copy Editor: Liz Lemal
Designer: Louise Leffler
Photographer: Kate Whitaker
Illustrator: Sarah Leuzzi
Prop Stylist: Nadine Tubbs
Production: Lisa Pinnell

Printed and bound in China

10 9 8 7 6 5 4 3 2 1

For my four favorite people:
Reuben, Ava, Oscar and Archie.

Contents

About this book

Science is completely and utterly brilliant. Not only does it make new discoveries, invent products, and save lives, but even more importantly it allows kids to get away with anything.

Yes, that's right—we're talking mess, mayhem, and even the occasional explosion—provided it's all "in the name of science", parents really don't mind. Well okay, if you blow up the kitchen they are going to be a *teensy* bit annoyed, but you get the idea: science = the perfect excuse.

Of course I should warn you that you are in danger of learning the odd thing along the way, but don't worry—it doesn't hurt. And just remember, the more you drop in words like **surface tension***, **friction***, and even **kinetic energy***, the less your parents are going to complain when they are wiping down the ceiling, mopping the floor, and wondering where half the contents of the kitchen cupboards have gone.

So pop on those safety goggles, don your white coat and let's get going— we have water balloons to launch, hovercrafts to build, ice-cream to make, helicopters to race, balls to levitate, rainbows to eat and a whole lot more chaos to create. But it's okay—it's science.

* You'll find proper, impressive-sounding, scientific words written in **bold**. If you're not sure what some of them mean, check the glossary on page 213, memorize the definition and then astound everyone with your knowledge.

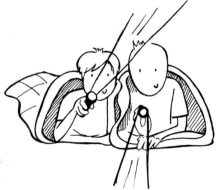

Booby trap a bottle

Air pressure is useful in lots of ways—inflating car tires, operating a vacuum cleaner, or flushing toilets—but by far the best use of air pressure is getting your friends soaking wet and then laughing at them. A lot.

First, you'll need a large bottle of water—this will work with a pre-filled bottle but equally you can booby trap the bottle and then add the water.

Whichever bottle you have, take your thumb tack, place the bottle on its side on a towel (with the lid on) and push the tack through in several places near the base. If your bottle is full of water this will start to leak through the holes, but it will stop when you stand the bottle upright. If your bottle is empty then once you've added your holes stand it in a sink and fill it right to the top. Although water will spray out of the holes, this will stop once you screw the lid back on tightly.

Now lift your bottle by its cap and be careful not to squeeze it as you place it on a table for your unsuspecting victim. Just remember, if you're lying in wait for so long that you start to get thirsty, DO NOT be tempted to have a swig from the bottle.

If you haven't caused enough mayhem yet:

Try adding different numbers, placement, and sizes of holes—what difference does this make? Does it alter things if you use carbonated (fizzy) liquid?

The Sciencey Bit

Air pressure is a **force** which pushes in all directions. When the lid is on, the air at the sides of the container pushes *in* the same amount as the water in the bottle pushes *out*, which means the water stays put. When you take off the lid, the air above is pushing down on the water in the bottle which forces it out of the holes at the sides—with hilarious but slightly damp results!

YOU WILL NEED: WATER BOTTLE WITH LID, TOWEL, WATER, THUMB TACK

Build a balloon-powered car

What do you mean "That'll be a fun way to travel to school"? This isn't a car for you to ride in. Then again, if we used enough balloons, you never know.

But first things first, let's see if we can get a *mini* car powered by a balloon.

You'll need to start with a base. Card works well for this as it should be stiff, but make sure it's also quite light—the less weight, the easier it is to move. Attach two straws to the underside with sticky tape, making sure they stick out farther than the card edge, and then thread through your kebab skewers—these are your car's axles.

Now to make your wheels. Trace around the cup to make four circles on corrugated card and one on paper and then carefully cut these out—the more perfect the circle, the better they will work. To cut down the **friction** caused by the rough edges of the card, you can add sticky tape so the shiny side provides a smooth surface for the wheels.

Take your paper circle and fold it in half and then quarters, then unfold and you will have located the center point. Lay this over your card wheels and push through a sharp pencil or pen to make the center holes before threading these onto either end of your car axles.

To create your engine, stretch the balloon out and then blow it up a couple of times so it's easier to re-inflate. Push the end of a straw into the neck of the balloon and attach it with duct tape. You may have to test and add more tape if there are any gaps—it must be airtight.

Finally, tape your straw onto the car so the balloon sits just on top of the card at one end and the straw sticks out over the other end. Place your car on a smooth surface, blow up the balloon through the straw, then let go. As the air escapes from the straw you should see your car move along.

If you haven't caused enough mayhem yet:

What about using a wider straw or a larger balloon—what difference does this make? Why not create your own design—you could try using a plastic drinks bottle for the body and bottle tops or cotton reels for wheels. Just make careful notes of which design goes fastest and furthest as you test.

 ## The Sciencey Bit

Congratulations! You have just demonstrated **Newton's Third Law of Motion**. This is a rule that says that for every action there is an equal and opposite reaction. In this case it means as the air rushes out of the straw it causes **propulsion**, which moves the car. It is also converting the **potential energy** (stored energy) in the elastic of the stretched balloon into **kinetic energy** (movement energy) of the car.

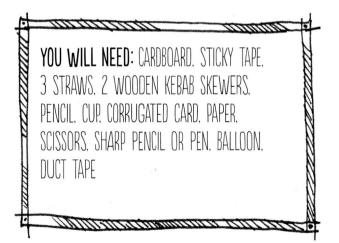

YOU WILL NEED: CARDBOARD, STICKY TAPE, 3 STRAWS, 2 WOODEN KEBAB SKEWERS, PENCIL, CUP, CORRUGATED CARD, PAPER, SCISSORS, SHARP PENCIL OR PEN, BALLOON, DUCT TAPE

Make a gravity-powered water siphon fountain

Why make this? Because when your parents ask you "What are you going to do today?" you can say "I'm making a **gravity**-powered water **siphon** fountain"—and watch the look on their faces. Seriously, it's totally worth it.

To begin this impressive-sounding project you'll need a small plastic bottle. Take off the lid, push sticky tack inside the cap and then use something pointed like an old pen or the end of scissors twisted round and round on top of the cap to make two holes (you will need to get your parents to stop gawping at your genius and oversee this bit as it can be a little tricky).

When the holes are large enough, remove the sticky tack and push a straw through each space—one straw should stick out about 4 inches below the lid, the other about 1½ inches. Now take your sticky tack and carefully mold it around the joins between the straw and the top of the lid so all gaps are sealed.

Before replacing the lid, add water to the bottle—it needs to be deep enough to cover the shorter straw end when the bottle is turned upside down.

Fill a glass with colored water and raise it up on some sort of object (a stack of books for example) before placing an empty glass of the same size on the surface below. Cover the end of the long protruding straw with your thumb and tip your bottle upside down so the short straw end is sitting in the colored **liquid** and the long straw end is positioned over the empty glass.

Now, take your thumb off the end of the straw and watch!

You will see water running into the lower cup and then colored water rushing into the bottle—or gravity pump—causing a fountain of water to erupt inside this. You will also probably see your parents staring in amazement at your brilliance.

If you haven't caused enough mayhem yet:

Try taping straws together to get a longer siphon, and then raise the top glass higher—what effect does this have? Why not set up siphons at different heights and race to see whose upper glass empties first?

The Sciencey Bit

The **force** of **gravity** pulls the water from the **siphon** pump into the lower glass. In doing this it lowers the **air pressure** in the bottle which pulls up water from the higher glass. The higher the top glass, the greater the effect of gravity and the higher and faster the water shoots in the "fountain".

Making the fountain

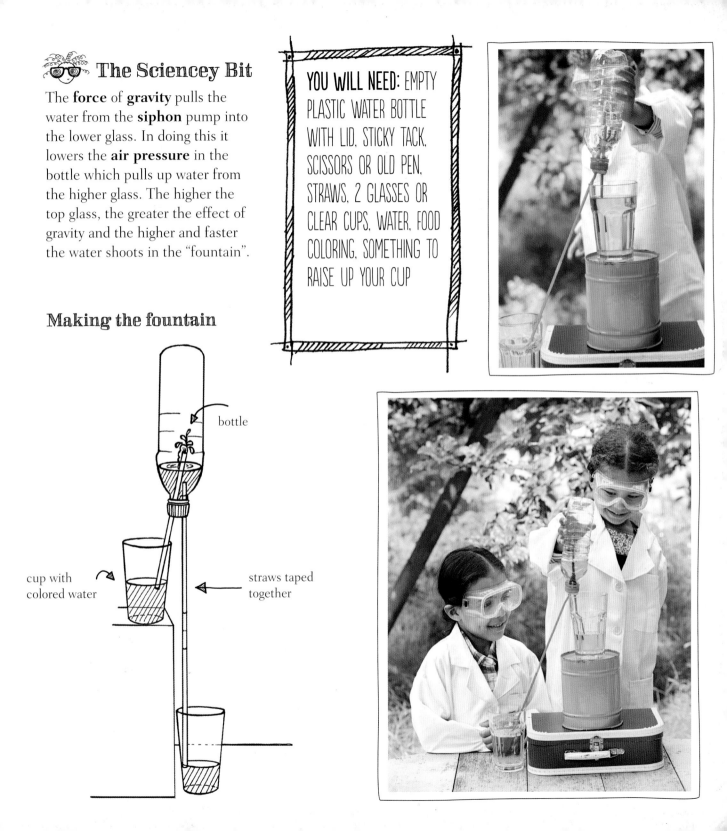

bottle

cup with colored water

straws taped together

Tip: Choose windows or glass doors that get lots of sunshine so your pictures really "glow".

Create a stained glass window

Poor windows, not only do they have the same view all year round but when anyone comes near, they simply stare straight through them. And really, there is nothing worse than being ignored—it's just so rude!

So maybe it's time to make people take notice of windows—and what better way than adding a bit of stained glass glamor?

To copy the look of the lead edging of stained glass, use a thick black wax crayon to make your pattern or picture on a sheet of paper. Go over the lines lots of times for a dramatic outline.

Now you can add your color. If you use the three primary colors of red, blue and yellow, you can then mix these to make all the other colors you may need (see Eat a Rainbow on page 120).

Paint each section in a different shade until all the areas are filled with bright colors. Try to keep inside the black wax crayon lines—after all, your windows are counting on you to do a good job.

When the painting is dry, turn it over so you are looking at the plain white back of the paper and begin to paint this with oil. Almost immediately you will see something amazing happening: what was once a blank white sheet is now magically displaying your colorful picture. When you're done, leave it to dry and then press it between two sheets of paper towel to mop up any oily bits.

Finally, use sticky tape to attach it to your window and watch as the sun shines through your masterpiece, giving the effect of stained glass. Right—try to ignore that window NOW!

If you haven't caused enough mayhem yet:

You could paint several pictures this way until the window becomes a stained glass art gallery. In fact, before you know it that pane of glass will have so many visitors it might start wishing for a quieter life.

The Sciencey Bit

Paper **fibers** are actually **transparent** (see-through), but when light is trying to pass through them they **refract** it (changing its direction slightly) so it cannot travel in a straight line. This makes the paper appear **opaque** (you can't see through it) because it has **absorbed** the light. When you paint it with oil, the oil fills up the air gaps and allows *some* of the light to pass straight through the paper, making it look semi-transparent—or **translucent**.

Blow a bubble snake

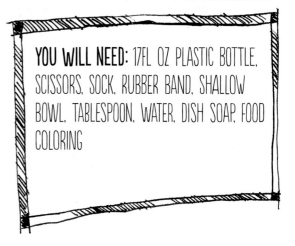

YOU WILL NEED: 17FL OZ PLASTIC BOTTLE, SCISSORS, SOCK, RUBBER BAND, SHALLOW BOWL, TABLESPOON, WATER, DISH SOAP, FOOD COLORING

Science can explain many things, but it is still baffled by the amount of odd socks in the world. Seriously, where do the other ones go?

Thankfully, you now have the perfect use for this mountain of odd socks—making bubble snake blowers.

Begin by cutting the base off a clean and empty 17fl oz bottle. Do this by taking off the top, squashing it flat and making a slit near the base. You can now re-inflate the bottle and carefully cut around the rest of it to remove the bottom.

Next, slip an old sock over the open end. This should be a *clean* sock—a snake is okay but the idea of one of your stinky unwashed socks … now that is terrifying. Stretch the sock nice and tight and then secure it in place with a rubber band.

Mix up a bubble **solution** in a shallow bowl using three tablespoons of water and one tablespoon of dish soap before adding a splash of food coloring.

Now take everything outside and dip the sock end of your bottle in the bubble mixture, take it out and blow. You'll immediately see a colorful snake start to appear. As the bubbles stick together you can keep blowing and blowing until the bubble mixture is used up and the snake is as long as you can make it.

And if you can find another odd sock you could even make a bubble blower for a friend. What do you mean, you have enough to make them for your entire class? Where have the other socks gone? Come on science—we need answers!

If you haven't caused enough mayhem yet:

You can try making a rainbow snake by dribbling lots of different food colorings onto the bubble mixture-soaked sock end before blowing.

The Sciencey Bit

When you blow air into the bottle, it forces the bubble mixture through all the tiny holes in the sock, making little bubbles. Bubbles like to minimize their **surface area,** so they join together to share common walls, making a huge snake.

Tip: Why not challenge a friend to see who can blow the longest bubble snake?

Make a nesting dolls set

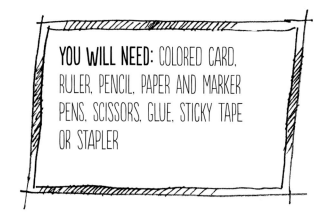

Adults say some weird stuff sometimes. They might tell you, for example, that "It's a dog-eat-dog world out there". If you stumble across such a deluded person, sit them down and explain that this makes absolutely no sense as a **food chain**.

If the poor person still doesn't get it, you can put together a simple food chain in a way even adults will understand.

Begin by deciding the food chain you want to demonstrate. You could start with the sun as it's the source of our **energy**, then go on to some sort of plant because these capture the sun's energy to produce food (which is also why they are called **producers**—see, it's so straightforward I reckon even an adult might get it).

The plant is then eaten—or *consumed*—by an animal—making it a **primary consumer**. But sadly for primary consumers they can also be eaten by **secondary consumers**. And before those secondary consumers get too smug, they should remember there is often someone else higher up the food chain—those pesky **tertiary consumers**.

When you have decided on the elements in your food chain, use colored card to make a rectangle for each one, progressively getting bigger as they move up the chain (see diagram).

Now you can decorate each food chain step by using pieces of paper glued on or drawing in details with marker pens. Remember to keep any drawings of animals, plants or even the sun to the middle third of the card so they can be viewed from the front when the rectangles are rolled into tubes.

When your rectangle is decorated, mark a line ½ inch in from the edge of your card, overlap along this line and sticky tape or staple your tube together.

Finally, stand your food chain up in a row and slot one over the other, while explaining to the adults in simple terms how dogs might eat ducks but definitely not dogs. Of course, you may have to do this a couple more times before they *fully* understand—they can be a little slow sometimes.

If you haven't caused enough mayhem yet:

Try making alternative food chains and see which is the longest one you can create.

The Sciencey Bit

All living things need food for survival. A **food chain** shows how plants, animals and humans rely on each other for food. All chains start with plants (also called **producers** in food chains) that capture **energy** from the sun. These are eaten by **herbivores** (plant-eating animals) which then receive the energy. The herbivores are in turn eaten by **carnivores** (meat eaters) who get the energy from the herbivores' bodies. Animals eaten by other animals are called **prey** while the animals that eat them are called **predators**. Adults who are confused about this are called "very silly".

Making the nesting dolls

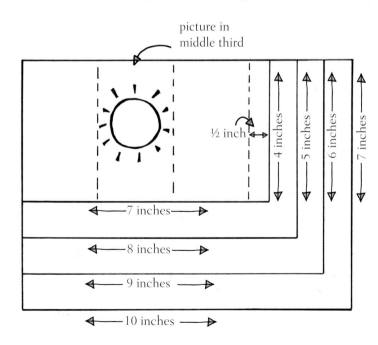

picture in middle third

½ inch

4 inches
5 inches
6 inches
7 inches

7 inches
8 inches
9 inches
10 inches

Launch a rocket

This is a great example of an **acid-base reaction**, a brilliant way to show the effect of temperature, and a clever way to synthesize **carbon dioxide**. But forget all those things—look at the title—we're going to *launch a rocket*—how cool is that?

Take the cap off your bottle and then use tape to attach four pencils, all the same size, to the outside (see diagram). Just make sure they are evenly spaced out with their ends lined up so the rocket doesn't … well … rock when you place it on the ground.

Pour the vinegar and warm water into the bottle, then cut a piece of paper towel in half and add the baking soda to the center before folding in the ends and rolling it into a thin sausage shape. Take these outside, along with a cork, which you need to wrap with a small piece of paper towel until you know it will fit snugly in the top of the bottle.

Right, now find a space, well away from anything and anyone else. At this point you will want to don your safety goggles and enlist the help of an adult. Carefully push your paper towel sausage into the bottle and then very quickly push in the cork, turn it upside down so it's resting on the pencil ends and RUN!

Within seconds the acidic vinegar and baking soda will have made a **chemical reaction** and formed a **gas** which escapes from the bottle with such a force it will launch your rocket into the sky. And if anyone tells you "That's not rocket science", you need to correct them.

If you haven't caused enough mayhem yet:

Try altering the amount of vinegar or baking soda and record any changes this makes to the way the rocket launches. You can also try using cold water—what difference does this make?

The Sciencey Bit

Baking soda contains something called a **carbonate**—or a base—which reacts with acids such as vinegar. One of the products of this reaction is **carbon dioxide gas** and as this builds up in the bottle it increases **air pressure,** which eventually forces the cork out of the bottle and pushes the rocket upwards.

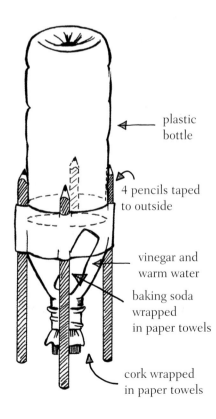

Making the rocket

Tip: Make sure there is nothing above where the rocket is going to be launched.

plastic bottle

4 pencils taped to outside

vinegar and warm water

baking soda wrapped in paper towels

cork wrapped in paper towels

Make ice cream

That's right—there is science in making ice cream. Admittedly, there is less science in *eating* ice cream but it is an unavoidable consequence so you'd best just be brave and get on with it.

Begin by whisking up your cream. If you use a hand whisk for this it should take you (or an unsuspecting adult who is forced to do your bidding) about five minutes, but it will be even quicker if you use an electric whisk.

When the cream has become nice and thick, you can sieve in the powdered sugar, then add the vanilla extract and raspberries. Mix everything together well and squish and break up the raspberries as you go along to add a nice pink color to the mixture.

When you're happy with how it looks, spoon it into a large shallow plastic container or even better, three or four small ones. Now put the lids on, place them in the freezer and set a timer for 45 minutes. When the time is up, take out the containers and use a spoon or fork to stir up the ice cream thoroughly. Put it back in and then do the same again 45 minutes later, and again 45 minutes after that. Finally, after the ice cream has been in the freezer for about 2 hours it should be hard enough to eat.

If you haven't caused enough mayhem yet:

Get experimenting! There are so many flavors and ingredients you can try adding to your cream and sugar base. How about blueberries, or strawberries, or bananas, or chocolate chips or mint, or all of them at once! Go on—you know you want to!

The Sciencey Bit

Whipping adds air to the cream, which is then trapped by the fat **molecules**—this makes the cream into a thick foam. When you freeze it, the water content in the newly frozen ice cream will form **crystals** and the smaller the ice crystals, the smoother the ice cream, so you need to stop them growing by freezing it quickly *and* breaking them up at regular intervals.

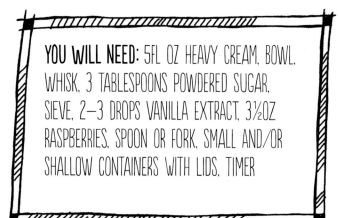

YOU WILL NEED: 5FL OZ HEAVY CREAM, BOWL, WHISK, 3 TABLESPOONS POWDERED SUGAR, SIEVE, 2–3 DROPS VANILLA EXTRACT, 3½OZ RASPBERRIES, SPOON OR FORK, SMALL AND/OR SHALLOW CONTAINERS WITH LIDS, TIMER

Tip: Homemade ice cream will keep in the freezer for a couple of weeks but is best eaten straight away... which is lucky!

Throw a boomerang

It wouldn't be great if most things you threw came back—javelins, darts, trash for example. However, boomerangs are different—they're supposed to come back … with practice anyway.

Traditional boomerangs were whittled from wood, but that would involve sharp knives and lots of patience, so I think we'd better stick to card.

First, measure out and cut three pieces of card 1½ inches wide and 6 inches long. You can use an old cereal box for this, or a postcard is the ideal size from which to cut all three pieces. Next, cut a 1-inch-deep slit in the base of each. Use this to slot the first two pieces together, and then the third (see diagram).

Use a protractor (yes, I know it's a math tool, but you don't need to look so worried) to make sure the angle between each of your blades is exactly 120 degrees—this means the wings are evenly spread. Hold them tightly between your finger and thumb while you check and then, when you're happy, staple them two or three times in the middle to hold them firm.

Now for a flying lesson. Hold one of the blades between your finger and thumb with the boomerang facing away from you. Set it off spinning by throwing it forward and up with a flick of the wrist. You should see it spinning around but following a curving path that will eventually bring it back towards you.

If you have a large room you can practice this inside where there will be no wind to add more complications. Then, when you've become a bit of an expert, take the boomerang outside and give it a spin (literally).

If you haven't caused enough mayhem yet:

See if you can make your boomerang fly in a different direction, or a larger circle. What happens if you try to throw it like a Frisbee? Try making different versions using alternative materials or by changing the shape of the wings.

The Sciencey Bit

As the boomerang spins through the air, the wing at the top of the spin is moving through the air at a faster rate than the wing at the bottom because it is moving in the direction you threw it. The result is that the top portion will generate more **lift** than the bottom portion as it cuts through the air. This difference in **forces** causes the boomerang to follow a curving path which should eventually bring it back to where it began.

Making the boomerang

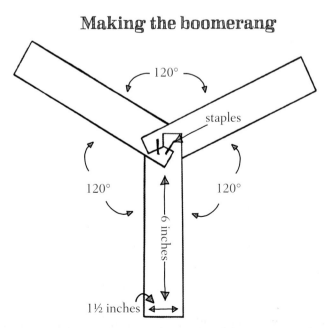

120°

staples

120°   120°

6 inches

1½ inches

Shoot an air cannon

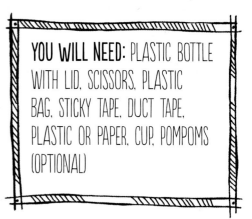

YOU WILL NEED: PLASTIC BOTTLE WITH LID, SCISSORS, PLASTIC BAG, STICKY TAPE, DUCT TAPE, PLASTIC OR PAPER, CUP, POMPOMS (OPTIONAL)

If you've ever wanted to shoot something off your brother's head (and let's face it, who hasn't?), this is a safer option than a bow and arrow and will definitely get you in less trouble.

Take your plastic bottle, remove the lid, squash it flat and make a cut near the bottom. Reshape your bottle and then use your scissors to finish cutting off the bottom section.

Use about eight pieces of sticky tape to hold your plastic bag in place at the end of the bottle then wrap around duct tape 2–3 times to make sure the two are sealed together with no gaps (see diagram).

Finally, blow into your bottle to inflate the bag and place this under your arm (but don't squeeze it!). Now, point the mouth of the bottle at the cup on top of your brother's head where it is balanced, and quickly move your arm to flatten the bag. A gust of air should shoot out of the air cannon and blow the cup right off his head.

And if you'd rather see some missiles whizzing through the air, you can pop some pompoms into the neck of the bottle and fire these instead.

If you want to watch something fly even farther, choose a pompom large enough to wedge in the top of the bottle and then shoot the cannon.

If you haven't caused enough mayhem yet:

Try adapting the design using larger or smaller bags or different-sized bottles—which work best? You could also try shooting your cannon at longer distances—or use it to blow out candles (with a bit of adult supervision, of course).

The Sciencey Bit

The air filling your bag and bottle is released when you squeeze with your arm, pushing air out in front of you and towards the cup. If you are in line with the cup, this **air pressure** will push it over. If you plug the end of your bottle with a pompom, the air pressure builds up behind it until eventually the pompom pushes it out of the way and it shoots forward.

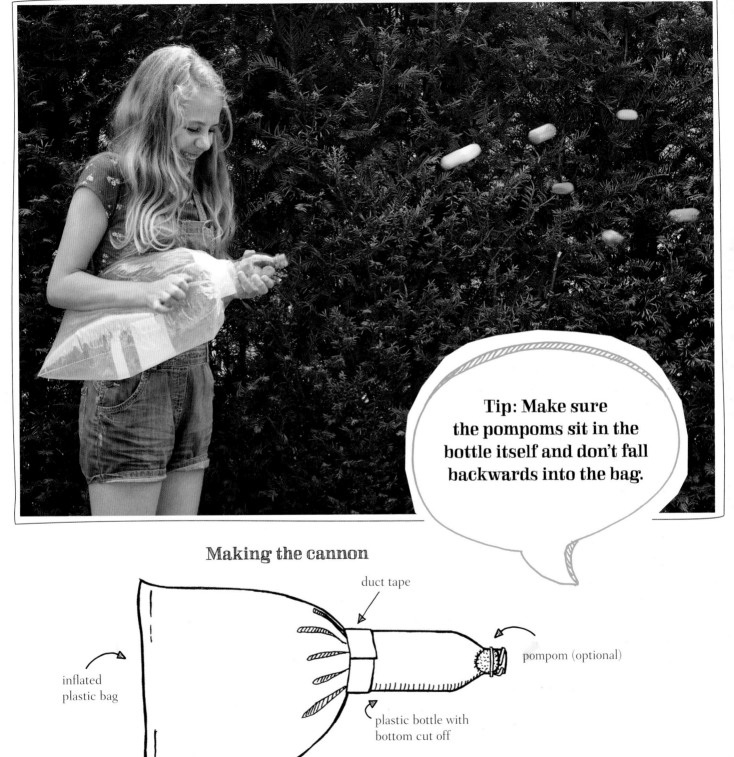

Tip: Make sure the pompoms sit in the bottle itself and don't fall backwards into the bag.

Making the cannon

duct tape

inflated plastic bag

pompom (optional)

plastic bottle with bottom cut off

Play the sticky rice trick

Anyone who struggles to pick up food with chopsticks is going to be super impressed with you after this trick. Never mind a mouthful or two, you are going to pick up an entire bottle of rice *and* you'll only need a single chopstick to do it. Amazing!

First, you need to prepare your bottles. Fill them both to the top with long-grained rice. The easiest way is to put the rice into a measuring jug and then pour it, very slowly and carefully, into the bottle. If you do this over a clean piece of paper any spilt rice can collect here and then be poured back into the jug.

Now take one of the bottles and begin tapping its base gently on a table so the rice packs down and you can add more to the space created at the top. Keep doing this again and again until there is no more space left and the rice is up to the top of the bottle.

Now you are ready to perform to your adoring audience (or just the cat if everyone's busy). Explain that you have two identical bottles filled with rice but that you have been training one batch of rice for many weeks to become strong in mind and body—so strong that they are now able to hold onto a single chopstick with incredible

power (don't worry if the cat is looking a little dubious at this point—everyone knows cats are highly suspicious creatures).

To prove your point, place your chopstick in the "untrained" rice bottle (the loosely packed one) and you will find it pulls out straight away. Now plunge a chopstick into your super-trained (packed) rice (you may need to wriggle and push it down quite hard to get it in) and watch as these grains hold onto it with such strength you can lift it high into the air, bottle and all.

If the cat could applaud I'm sure it would.

If you haven't caused enough mayhem yet:

Experiment with different sizes and shapes of plastic bottles—which works best and why do you think that is?

 **The Sciencey Bit**

The air spaces between the rice in the first bottle allow the grains to pass easily past each other—and the chopstick. However, as you pack the rice tightly in the second bottle there is less room for the grains to move and more of them rub against the chopstick, causing **friction**—when this **force** of friction becomes **greater** than the force of **gravity** pulling down on the rice and bottle, you can lift them in the air.

YOU WILL NEED: 2 CLEAN, DRY, IDENTICAL PLASTIC BOTTLES, LONG-GRAINED RICE (E.G. BASMATI), MEASURING JUG, PAPER, CHOPSTICKS

Tip: Try using different objects for the rice to cling to, such as a pencil, a butter knife or even a wand.

Make fizzy sherbet

Yes—you read that right. We are pushing forward our knowledge of science and testing **chemical reactions** by … making candy. We even get to call a lollipop "an essential piece of equipment". Happy days, kids, happy days!

In fact, why can't all learning work this way? It would do wonders for everyone's test scores.

Begin by placing the baking soda in a bowl, then adding your citric acid, sugar and jelly crystals and stirring them all together. If you don't have jelly crystals, you can add a drop or two of food coloring and flavoring to the bicarbonate of soda, mixing these in well before adding the other ingredients.

Now lick your lollipop (I'm guessing you don't need detailed instructions for this), dip it into the sherbet mixture and lick it again.

You should notice it fizzing on your tongue. Do you know what that is? Yes, yes, delicious, but also a chemical reaction. What's that you say? You need to retest your findings to be sure? And again? Wow—your dedication to science is quite exemplary.

If you haven't caused enough mayhem yet:

Try tweaking the recipe to see if you can make more fizz—or a different flavor. Just make sure you only add small amounts of ingredients at a time and keep noting down the changes after each addition so you can keep track of what works best.

 The Sciencey Bit

The baking soda (an **alkali**) reacts with the citric acid (an **acid**) but only when water is involved—this is helpfully provided by the saliva in your mouth. The reaction produces a **gas—carbon dioxide**—which also creates the bubbling, fizzing sensation.

YOU WILL NEED: BOWL,
2 TEASPOONS BAKING SODA,
1 TEASPOON FOOD GRADE
CITRIC ACID,
4 TEASPOONS POWDERED
SUGAR OR SUPERFINE
SUGAR, 2 TEASPOONS
FLAVORED JELLY CRYSTALS
(OPTIONAL), FOOD COLORING
(OPTIONAL), FLAVORINGS
(OPTIONAL), LOLLIPOP (NOT
OPTIONAL—IF PARENTS ARE
READING THIS)

Tip: Wearing swimsuits will make life easier if this trick doesn't go to plan.

Tip water over someone's head

Please tell me you have read further than the title—otherwise you and this book may be parting company very soon. Yes, there is a bit more to this than simply tipping water over people's heads—otherwise why do you think I've written all these other words?

In fact, this is a very effective way to demonstrate **air pressure** while impressing your friends. First, you'll need an assistant—preferably someone who won't be able to ban you from TV for a month if things go a little wrong and they get soaking wet.

Now fill a glass or strong plastic cup with water, right up to the rim. Place your card on top, tapping it in the middle to make sure a seal has formed between the card and water. Holding the card in place, tip the glass upside down until it is hovering over your assistant's head, then … move your hand away from the card. If all goes to plan, the card will stay in place and the water remains in the glass. If it doesn't go to plan … run away … quickly!

If you haven't caused enough mayhem yet:

You could try the **experiment** with bigger containers—but make sure you do this outside where a huge spillage of water will not cause *too* much damage.

The Sciencey Bit

Air pressure is a **force** and because air is a **gas** this force not only pushes *downwards* but in all directions—including *upwards*. The card stays in place because the pressure from the air pushing upwards is greater than the pressure of the water pushing down.

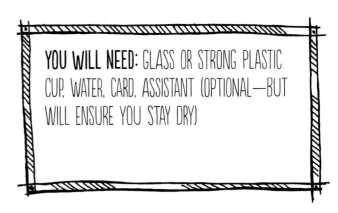

YOU WILL NEED: GLASS OR STRONG PLASTIC CUP, WATER, CARD, ASSISTANT (OPTIONAL—BUT WILL ENSURE YOU STAY DRY)

Build an aquarium

If you want to keep fish but you're a bit … well … lazy, then this is the answer: no feeding, no tank cleaning, no plant checks, no water changes. Even better, when one of your fish is floating on the surface, there's no need to panic.

In fact, you want your fish to **float,** which is why you'll be making them out of craft foam. Begin by drawing fish shapes on the foam, then cut these out and use a hole punch to add eyes. You can also use foam to cut out seaweed shapes; just check them against the size of your chosen aquarium bowl to make sure they're not too big.

Add a layer of sand or gravel—and one or two artistically arranged shells—to the base of the bowl. Next, pass a piece of cotton thread through one of your fish's eyes then wrap both ends of the thread around a small rock before placing this on the aquarium floor. You can secure your seaweed in a similar way by punching a hole at the bottom and passing cotton thread through. Make sure that the entire length of thread is wound around the rock so your seaweed doesn't float up.

Now slowly pour in water until it reaches nearly to the top of your bowl and watch what happens. As the sand settles, you should be able to see your fish floating happily in the water. If any are too low, you can adjust them by unwinding more cotton from the rock. And if any are floating on the surface then we need to shed a tear or two, say a few words of remembrance for dear Nemo and get ready for the flush of fate.

Oh no, sorry, got a bit carried away there. Actually, you can just wind a bit more cotton thread onto the rock until it gets lower in the water. It's okay, you can stop crying now.

If you haven't caused enough mayhem yet:

See if you can find any other materials which will make floating fish. Can you work out how to make a fish hover horizontally in the water like a manta ray?

The Sciencey Bit

As well as building a beautiful aquarium, this project can help you explore **density** through some floating and sinking **experiments**. Stones will sink to the bottom of the aquarium because they are heavy for their size—this is called their density. Sand will also sink as it's more dense than water, but because it's also very fine it starts by forming a **suspension** (which makes the water look cloudy) before settling. Because craft foam is full of trapped air, it is far less dense and will float to the top (unless you tether it with cotton).

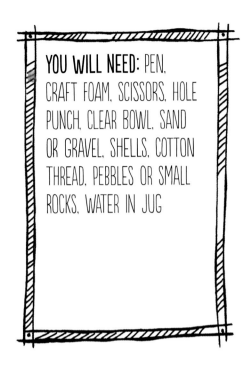

YOU WILL NEED: PEN, CRAFT FOAM, SCISSORS, HOLE PUNCH, CLEAR BOWL, SAND OR GRAVEL, SHELLS, COTTON THREAD, PEBBLES OR SMALL ROCKS, WATER IN JUG

Tip: You can also secure your seaweed by trapping the bottom of the foam shape between a rock and the bowl.

Spin a centrifuge sprinkler

YOU WILL NEED: STRAW, THIN WOODEN KEBAB SKEWER, RULER, PEN, SCISSORS, STICKY TAPE, JUG OR BOWL, WATER

You know what's perfect for a hot sunny day?

Well yes, you're right, ice cream is perfect. But do you know what *else* is perfect?

Okay, fair enough, ice lollies are also perfect. But do you know what else is perfect?

Centrifuge sprinklers! (Um, well, no, you can't eat them so let's just call them "nearly perfect" instead).

To make a sprinkler, take a good-sized straight straw, bend it in half to find the midway point, then carefully poke the kebab skewer through this and push it a couple of inches through. Now measure 1¼ inches in either direction from this hole and cut halfway through the straw at each point. Bend both sides of the straw to make a triangle shape (see diagram) and tape these in place (without covering the ends of the straw).

Finally, place the pointed end of the triangle shape into a jug or bowl of water and twist the top of the skewer. You should find that water is flung in all directions.

Ah, yes. I probably should have told you to go outside first. Sorry about that.

If you haven't caused enough mayhem yet:

Try using different sizes and lengths of straws—how far can you make the water travel? How many different people can you soak in one spray?

The Sciencey Bit

If an object is spinning, anything on that object appears to experience a **force** pushing it outwards. This is called **centrifugal force**. In your straw, the water is pushed outwards and as the only way it can travel in that direction is to move up the straw it also means it is pumped upwards. If you spin the straw fast enough, water will exit the straw and fly outwards as it is now moving quite fast.

Making the sprinkler

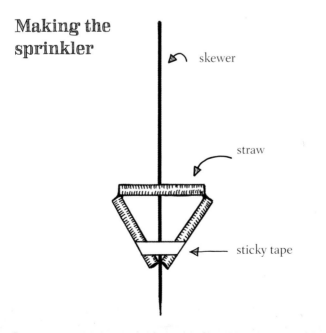

skewer

straw

sticky tape

Make bubble wands

No, you can't turn your brother into a frog. Or a newt. Or a newt's bottom. Yes, these are wands but I never said they were magic, did I? Still, they can turn any shape into a sphere so that's something, isn't it?

First, take your twig and wrap a couple of inches of the wire around one end. Then take the other end of the wire and wrap a couple of inches of this over the first bit of wire. You should be left with both ends of wire securely attached to your twig wand and a large loop of wire between them.

Next, select a shape—cookie cutters work well, but you could look at any object that will form a good mold (no, NOT your brother—put the poor boy down).

When you have one, place it in the middle of the loop of wire which you can then shape around it. Try to make sure there is the same amount of wire left over at both sides and then twist these together to close up the shape before slipping out your cookie cutter or mold.

Finally, mix up your bubble **solution** (one part dish soap to three parts water), dip in your wand and blow.

Amazingly, whatever shape your wand is—even if it's the shape of a newt, or a newt's bottom—the bubbles will stubbornly ignore it and remain spherical. That is just what bubbles do.*

*Except when they are square, of course— see page 78.

If you haven't caused enough mayhem yet:

You could decorate your twig with acrylic paint or wrap it in yarn to give it a colorful look.

The Sciencey Bit

Whatever shape your wand is, the bubble that forms will always be spherical. This is because the stretchy skin of the bubble always wants to pull back into the smallest possible shape—a sphere.

Tip: For thicker, longer lasting bubbles, you can add a few drops of glycerin to the mixture.

Plant a miniature desert garden

Poor plants! If a place is very hot, very cold or just downright tricky, they can't pack their bags and move. Instead they have to learn to live with it—or adapt.

Thankfully, plants are very good at adapting, especially the desert dwellers. And if you want to see these clever plants in action, then why not create a mini desert of your own?

First, fill your container with "desert soil". The water must run through it easily, so make it by mixing two parts of potting soil to one part grit or fine gravel and then place this in your pot, pressing it down gently as you go.

And now it's time for the plants. You can use cacti or succulents or a mixture of both.

If you're planting cacti, be careful! One way they have adapted to the desert is by having sharp spines instead of leaves. You can adapt to this yourself by using folded strips of newspaper as planting tongs or even putting on oven gloves. You'll look a bit odd but it'll be a lot less "ouchy".

Succulents are altogether friendlier. They have a waxy, almost rubbery feel to their leaves which stops water **evaporating** too easily.

Whichever you choose, dig a hole, place in the plant and then fill back around it with your potting mix, pressing it down firmly. Finally, give everything a good watering and let all the extra liquid drain away.

To get the authentic desert look you can add a shallow layer of sand to the top and maybe scatter a few props around too—scorpions, snakes, small Bedouin villages, whatever takes your fancy.

Water your garden by standing the container in a basin of water for an hour once every 2–3 weeks from spring to fall—then don't water it at all over winter. And make sure you place your garden somewhere very bright and sunny in the house. Oh, and finally—most important of all—however much you love your desert garden, *never* be tempted to hug a cactus!

If you haven't caused enough mayhem yet:

Try planting succulents or cacti in lots of different containers. As long as you can add holes to the base, use your desert mix soil and keep them somewhere sunny, you can be as imaginative as you like—old shoes, toy trucks, cans, it's up to you.

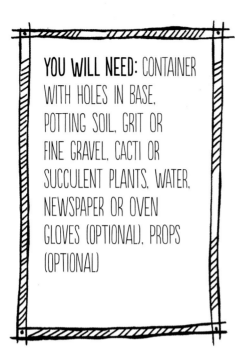

YOU WILL NEED: CONTAINER WITH HOLES IN BASE, POTTING SOIL, GRIT OR FINE GRAVEL, CACTI OR SUCCULENT PLANTS, WATER, NEWSPAPER OR OVEN GLOVES (OPTIONAL), PROPS (OPTIONAL)

 **The Sciencey Bit**

Plants such as cacti and succulents have adapted over millions of years to survive in harsh conditions. **Adaptations** such as storing water in their stems or having waxy cuticles (a protective layer) on leaves to stop water **evaporating**, mean these plants are able to grow in the hot and dry conditions you've recreated in this mini desert.

Tip: You can use toys and other small objects to liven up your desert landscape.

43

Release bathtime fountains

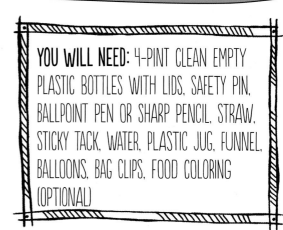
If you want to bathe with a bit more splendor, how about setting up a cascade of fountains? Let's face it, you deserve nothing less.

For each fountain you'll need an empty 4-pint plastic bottle with a lid. Keep the lid screwed on while you make a hole in the side, about a third of the way up. It's easiest to start this off by pushing through the sharp end of a safety pin, then make it big enough for a straw by pushing and twisting through the nib of a ballpoint pen or sharp pencil.

Now slot in the straw and then mold your sticky tack around it, making sure the straw is angled upwards and there are no gaps between it and the bottle. If you have a bendy straw, you can point the end up, too.

Now take off the lid and use your jug and funnel to add water to the bottle. It's best to do this by the bath in case of spillages. You can fill the bottle in line with the end of the straw without water escaping because it will sit at the same level in both.

Now blow up a balloon, use a bag clip across its neck to trap in the air and then put the mouth of the balloon over the top of your bottle.

When you have set up all your bottles, run a nice bath, hop in and then quickly take off all the clips. Your fountains will start spouting water in arcs adding a suitably dramatic and regal feel to your bath. When they finish, you can refill the bottles, blow up the balloons and do it all again and again (or at least until the bath gets cold).

If you haven't caused enough mayhem yet:

Try adding different food coloring to the water in each bottle to make a rainbow of fountains, or try setting straws at different levels to see what happens.

The Sciencey Bit

The balloon is made of an elastic material that stretches to take in the air you have blown into it but will then return to its original size and shape. The air leaving the balloon enters the bottle, which is already full of air. The extra air increases the **air pressure** and this pushes down on the water below, causing it to shoot out of the straw.

Chop up a fruit salad

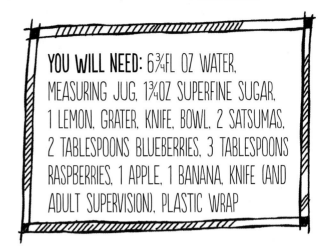

YOU WILL NEED: 6¾FL OZ WATER, MEASURING JUG, 1¾OZ SUPERFINE SUGAR, 1 LEMON, GRATER, KNIFE, BOWL, 2 SATSUMAS, 2 TABLESPOONS BLUEBERRIES, 3 TABLESPOONS RASPBERRIES, 1 APPLE, 1 BANANA, KNIFE (AND ADULT SUPERVISION), PLASTIC WRAP

Now, now, calm down. I know this has "fruit" *and* "salad" in the title, but despite those scary words it's actually a delicious pudding. And yes, you can have puddings that don't involve ice cream or chocolate. No, seriously, I'm not making that up.

Begin by mixing a syrup (see—that doesn't sound too healthy, does it?). Pour your water into the measuring jug then add the sugar (there you go—even better!) and stir it until all the sugar has **dissolved**, which means it doesn't separate out when the water stops moving.

Grate some lemon rind and add it to the syrup before cutting a lemon in half across the middle (you may need to ask an adult to do this) and squeezing the juice out of it—into a lemon squeezer if you have one, or if not into a bowl (taking out any pips which escape). Now add this juice to the water.

Peel your satsumas and separate the segments before placing them in the bowl along with the blueberries and raspberries.

Now comes the tricky part: cutting up your apple. You will need an adult to help. If they take a slice off one edge, the apple will then lie flat on a chopping board, making it easier to handle. Cut a couple of slices off each side and then lay these flat before chopping them into smaller pieces and placing in the bowl. Next, peel your banana and carefully chop it into slices before adding these to the rest of the fruit.

Finally, mix everything together well, pour in the syrup, put plastic wrap over the top of the bowl and leave it in the refrigerator to chill for a couple of hours. When you're ready to eat, remove the plastic wrap and serve. And yes, ice cream does go rather well with this.

If you haven't caused enough mayhem yet:

Leave pieces of different cut fruits out in the air to see which ones turn brown and how quickly—why do you think some take longer than others? How about if you put the pieces in the refrigerator? Try making the syrup using warm water and cold water—which one **dissolves** the sugar fastest? Try adding more ice cream to your bowl—how long before an adult spots what you're doing?

The Sciencey Bit

So why does some cut fruit turn brown in the first place? Well, oxygen in the air reacts with products in the cut fruit **cells** and one of the results of this is a change in color. The process is called **oxidation**, which is the same reaction that causes metal to rust. However, the chemicals that cause this to happen—called **enzymes**—can't work in very acidic conditions, so the **acid** in the lemon juice stops oxidation taking place, keeping your fruit salad looking fresh and delicious.

Make foil leaf art

When people say "Nature is the best artist", they are right. But what I say is, if nature's so good, why not steal a bit of her artwork and pass it off as your own? Cunning, eh?

First, you need to find some impressive fall leaves. That's right, we're going to do this with bits of artwork nature doesn't even want any more—that's quality recycling.

Stick your leaf, or a selection of leaves, to some cardboard, or even the center of a paper plate using white glue. Now spread another layer of white glue across the card or plate as well as the leaf and stick on some aluminum foil. You can choose the shiny or the dull side depending on which look you prefer. Oh, and don't worry if it's a little large, as you can trim it to size or simply fold it over the back out of sight.

Using your fingers, smooth out the foil and you should see the shape of the leaf emerging. In particular you will see the raised leaf **veins**.

You can leave your artwork like this or, if you want to create a two-tone look, add a couple of layers of black paint over the top. When these are fully dry, use a small piece of foil folded over a few times to make a rubbing pad about ¾–1¼ inch square. Mold this over your forefinger and use it to gently and carefully rub over the leaf's raised areas, especially the edges and veins.

You should see the paint start to come away in these places, making the leaf skeleton glint through the black paint. You could even use the pad to remove a little paint around the edges of your picture to add a silvery frame. And while you're at it, why not take the end of a chopstick or something similar to "sign" your artwork by scratching over the black. Maybe "Artwork by Max aged 9 and Nature aged … a lot more".

If you haven't caused enough mayhem yet:

Try using different colors or leaf shapes and see which creates the best effect. Can you spot differences in the patterns of the veins? Why not use leaf art to make cards or gift tags?

The Sciencey Bit

As well as creating a very stylish bit of artwork, this project means you can get a closer look at how leaves work. The **veins** are all over the leaf because they are made up of tubes that form the tree's transport system. Plants like trees use these tubes to move **water and minerals** to their leaves and to move sugars, made by **photosynthesis**, out of the leaf to other parts of the plant which need them.

Tip: You can try using different colors of paint to create a variety of looks for your foiled leaf art.

YOU WILL NEED: FALL LEAVES, PIECE OF CARDBOARD OR A PAPER PLATE, WHITE (PVA) GLUE, ALUMINUM FOIL, PAINTBRUSH, BLACK PAINT

Polka dot a lawn

YOU WILL NEED: STRING, PENCILS, OLD CARDBOARD BOXES, SCISSORS, STONES, LAWN

Lawns don't have a very exciting wardrobe, do they? It's kind of a choice of green, green or green. So how about adding in a few polka dots to give your turf a whole new look? Before you start, though, it's best to ask an adult for permission.

First, make some dots by drawing circles onto your old cardboard. To do this, take a length of string and tie it to a pencil near its pointed end before pushing this into the middle of your card. Now tie the end of the string to another pencil in the same way, making sure its length is half the **diameter** (length across the middle) of the circle you want to create. Move the second pencil around while holding the middle one still so they act as compasses allowing you to draw a circle. Do this as many times as you want.

You could create different-sized dots or all the same, just make sure they are at least 7 inches across.

Cut out your dots and then arrange them on the lawn. When you are happy with the way they look, weight them down with some stones so the card won't blow away.

And now you need to leave them for about a week (yes, I know, it takes lawns a very long time to get changed. Thank goodness they don't need to dress for school every day).

After the time is up, lift up a dot and see if you can spot a color change underneath. If it's not obvious enough, leave it for a couple more days and check again. When you're happy with the look, remove the card to reveal your green lawn with its very fetching yellow polka dots (which will gradually return to their usual green within a couple of weeks ... ah, shame!).

If you haven't caused enough mayhem yet:

Try using giant cardboard letters to write your name on the grass. After all, everyone loves wearing designer name labels—even lawns.

The Sciencey Bit

Lawns look green because the grass contains **chlorophyll**—a green **pigment** which plants use to trap the sun's **energy**, which is then converted into food. Chlorophyll breaks down easily and has to be constantly replaced. The grass beneath the cardboard isn't exposed to the sun, which means it has no reason to replace the chlorophyll so it loses its green color, turning yellow.

Stab a leak-proof bag

YOU WILL NEED:
SEALABLE FREEZER BAG, WATER, SHARP, ROUND-EDGED PENCILS

This is very much like the famous trick where you put a lady in a box and then stick swords through it.* Except there isn't a lady. Or a box. Or swords. But apart from that, it's very similar!

First, run plenty of water into your bag and seal it at the top. Now dangle the bag over your assistant's head and, with as much fanfare and build-up as you can manage, take one of your pencils and push it, sharp end first, into the bag and far enough through that the sharp point sticks out the other side.

Wait to hear the audible gasp from your audience as they realize the water is not pouring over your assistant's head but is still safe and sound in the bag. Okay, they might not exactly gasp, but I bet they'll still be quite impressed.

You can keep doing this again and again, until your audience are overwhelmed with your genius … or just get bored and leave.

*Please don't try the sword/box/lady trick at home. It won't end well.

If you haven't caused enough mayhem yet:

Experiment with different thicknesses and shapes of pencil and a range of bags and amounts of water—which make the perfect combination (and most impressive-looking trick)?

The Sciencey Bit

Your plastic bag is most likely made of a **polymer** called Low Density Polyethylene (LDPE) which consists of incredibly long **molecules**—a bit like strands of spaghetti. It's easy to push the sharp pencil between these molecules and they are flexible enough to form a temporary seal between the bag and the sides of the pencil. Remove them though and the water leaks out as the molecules have been pushed out of place permanently.

Tip: If you haven't got a sealable bag, you can use an ordinary freezer bag and tie a knot in the end.

Tip: Try to do this in
a wide open space
away from anything
breakable.

Super bounce a ball

Isn't it annoying when people are bigger, stronger and better at things than you? Tennis balls probably feel the same way when they look around at bigger, bouncier balls. But rather than get annoyed, they just borrow a bit of that **energy** for themselves. The sneaky so and sos!

If you want to see this genius in action, first take a tennis ball, hold it at shoulder height and let it drop on a hard surface to see how high it can bounce. The answer is probably "not that high"— sorry, tennis ball, the truth hurts.

Next, take the same tennis ball and hold it just as high but this time place it on top of a soccer ball before dropping them both together.

Now you will see the tennis ball bounce way higher than it did before, leaving the soccer ball far behind, and generally feeling very smug.

If you haven't caused enough mayhem yet:

Try this **experiment** with different-sized balls and see what difference it makes.

The Sciencey Bit

The tennis ball hits the ground, which pushes back with the same **force**, bouncing the ball up into the air. As the soccer ball is bigger and has a greater **mass**, it hits the ground harder, making the ground push it back with more force than the tennis ball. When they both hit the ground together, they are pushed back into the air with the force of their combined mass and when they separate, this causes the tennis ball to travel farther.

Weeee!

Make a giant naked egg

Ever wondered what an egg would look like without its shell? Well here's your chance to find out. Oh, and don't worry—the eggs don't mind. They aren't at all shy. In fact, they're total exhibitionists—or should I say eggs-hibitionists?

First, place your egg gently in a glass or jug and then pour in enough vinegar to cover it. As you watch, you'll see bubbles appearing all over the shell. If you want to impress people you can tell them the **acetic acid** in the vinegar is reacting with the **calcium carbonate** in the shell and releasing **carbon dioxide gas**. Then again, you could just say "Ooooh, look, bubbles!" and point.

Leave the egg in for a day, then rinse it off and replace with a fresh lot of vinegar. In a couple more days you should be able to carefully take your egg out and very gently rub off the final bits of shell. What you are left with is a naked egg surrounded only by a **membrane** (a thin layer of tissue) which means you can clearly see the yolk floating around inside.

You should notice that the egg is slightly larger now—that's because the shell is **semi-permeable,** which means it will let through *some* things, like water **molecules**. It also prefers things well balanced—as the vinegar is about 95% water and the inside of the egg is only about 90% water, it lets some extra water across until the concentrations are about the same.

In fact, you can make it even bigger. Just place the egg in a jug of pure water and leave it overnight, and it will take in even more water to try and balance things out. You could also try adding food coloring to the water when you resoak your naked egg because the only thing cooler than a giant naked egg, is a giant naked green egg!

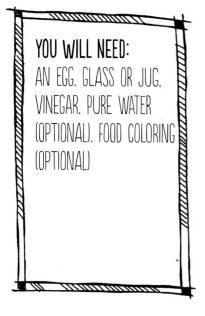

Tip: You can use different food coloring to make giant eggs in all sorts of shades.

If you haven't caused enough mayhem yet:

The naked egg is quite rubbery so you can have a go at bouncing it. No! Just from a few centimetres height. Now look at the mess you've made! Try the **experiment** using hard-boiled eggs—how does it compare to using a raw egg (apart from being less messy when you drop it, of course)?

The Sciencey Bit

You have just created a **chemical reaction** in which an **acid** (vinegar) reacted with an **alkali** (the **calcium carbonate** which forms the eggshell) to **corrode** the shell and release **carbon dioxide**. You have also seen **osmosis** at work, where water has passed through a **semi-permeable membrane** to make the concentrations either side equal. In fact, you've had a very busy day—I'd sit down if I were you.

Tip: It's easier to blow up the balloons if you stretch them lots before you start.

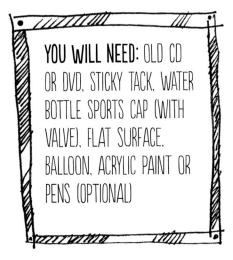

YOU WILL NEED: OLD CD OR DVD, STICKY TACK, WATER BOTTLE SPORTS CAP (WITH VALVE), FLAT SURFACE, BALLOON, ACRYLIC PAINT OR PENS (OPTIONAL)

Build a hovercraft

If you don't know what a hovercraft is, it's a type of craft that hovers. Good. Glad we've cleared that one up.

Of course, hovercrafts are usually designed to carry passengers. This one won't (just in case you were thinking of hopping aboard). But it still hovers and looks brilliant, so you should build it anyway.

First, you need to find an old CD or DVD that's *no longer wanted*. This is important because however much adults like you doing science experiments, they will *very* quickly go off the idea when you destroy their favorite DVD. And when you're *sure* it isn't wanted you can also add a bit of bling to the top of your disk with acrylic paint pens or glue-on decorations—after all, science and fashion aren't mutually exclusive.

Now take your sticky tack and stretch it about until it's nice and soft, before rolling it into a long sausage shape. Bend this into a circle and press it around the edge of the hole in the center of your CD or DVD making sure there are no gaps in the doughnut shape.

Remove the cap from the drinks bottle—push the valve down—and press this into the sticky tack so it's held tightly.

Place the disk on a flat, smooth surface—tables work well—before taking a deep breath and blowing up the balloon. When it's fully inflated, twist the neck around a few times to stop the air coming out too quickly and then hook the end over the bottle cap.

Finally, lift up the valve, give the disk a push and watch it race away! What do you mean, it's not moving? I thought I said "No passengers". Now take the cat off it and try again. See—that's better (and the cat looks less worried too).

If you haven't caused enough mayhem yet:

Try finding bigger balloons to use—after all, more air = more hover time. You can even try playing a short game of air hockey with you and a friend standing either end of a table trying to score by shooting the disk over the edge.

The Sciencey Bit

When objects rub against each other they produce **friction**—this is a **force** which resists motion and slows things down. If you just tried to push the CD along the surface without the balloon, there would be a lot of friction. Once you attach the balloon, though, and set your hovercraft in motion, air escapes from the balloon, forming a cushion of air between the disk and the table. This means there's hardly any friction so the disk can move far more quickly.

Make a jumping bean

You know when you're running towards lunch at full pelt then suddenly you realize boiled cabbage is on the menu? And you try to stop yourself but it's too late and you crash into the lunch lady who is not amused? Well that's called **momentum** (it's also called "really embarrassing", but we'll brush over that for now).

Momentum works for other things too—like making a jumping bean. In fact, this bean isn't a bean, and it doesn't exactly jump, but it is going to impress your friends a lot more than crashing into lunch ladies, so let's forgive it.

First, cut a piece of foil about four times as wide and five times as long as your marble (or around 2¾ inches x 4 inches). Roll it lengthways around your finger or a thick marker pen before placing the marble inside the foil tube you've created. It must have just enough room to roll along so adjust the tube if you need to.

Now keeping your marble inside, carefully seal the tube ends by pinching or folding them in, but be careful not to dent the tube itself or this will stop the marble moving about.

Put your foil "bean" carefully inside a plastic pot with a lid and shake it for about a minute.

This helps smooth out the ends of the bean by knocking them against the sides of the pot.

Remove the lid and you will see your finished "bean". Now you can tilt the tub back and forth to see the bean flipping about as though it's alive. You can also take the "bean" out of the container and pass it gently from one open palm to another to get the same effect—and distract any slightly angry looking lunch ladies.

If you haven't caused enough mayhem yet:

What happens if you try this with larger or smaller marbles, heavier or lighter balls? Why does foil work so well for the bean? And what other materials could you try? Which design is best for distracting lunch ladies?

👓 The Sciencey Bit

As you tilt the tub, the marble rolls along until it hits the end of the foil. Because it's traveling at speed in one direction (its **velocity**)—a bit like you hurtling at speed towards the lunch lady—it is hard to stop, and this **momentum** causes the foil bean to flip. The faster the velocity and the greater the **mass** of the marble, the larger this momentum will be.

YOU WILL NEED: ALUMINUM FOIL, A MARBLE, SCISSORS, PLASTIC POT WITH LID (TRANSPARENT IS BEST), THICK MARKER PEN (OPTIONAL)

Tip: For really glamorous jumping beans, try using colored foil.

Fly a carp kite windsock

Carp kites—or *koinobori*—are used to celebrate Children's Day in Japan. Of course, loading children down with their own bodyweight in chocolate might be a more obvious way to celebrate, but hey, these are nice too.

To make your own, take a sheet of tissue paper and fold it in half lengthways. Now use a pencil to draw on a carp shape (see diagrams). You can make the tail part of the body of the kite or else attach streamers for the tail afterwards.

Cut out the shape, then open it up and begin to decorate. You can stick on other pieces of tissue paper for the eyes and scales or perhaps just write on with pens—maybe "It's Children's Day … please give generously … chocolate always welcome".

Cut a strip of paper about 1¼ inch wide and glue this all along the mouth edge of the carp, then bend it over and stick it down again before punching three evenly spaced holes along its length. For the streamer version you can also **reinforce** the back of the fish in the same way.

Now stick together the two long edges of the fish (but leave the tail untouched). It can be easier to do this by placing one arm inside the fish to give you something to press against. If you are using a glue stick, be careful not to rip the tissue paper—using white glue applied with a small paintbrush may be easier.

Staple on streamers (if you are using them), then attach a 24-inch length of string through each of the mouth holes before tying the ends of these to a bamboo pole or stick. Push the end of this into the ground and wait for the wind to lift your carp.

If there isn't much wind, you can create your own **lift** by running with your kite held aloft. If you start to tire doing this, you may need to consume more chocolate—just for energy, you understand.

Making the carp

with tail

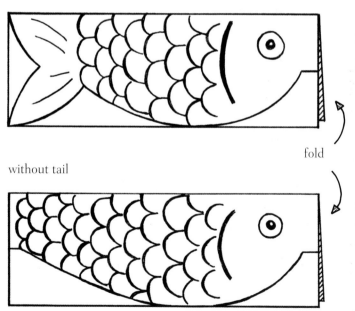

fold

without tail

If you haven't caused enough mayhem yet:

Try using different materials to make carp kites—what about card, or fabric, or plastic from carrier bags? Which work best and why? Does glue work or do you need to use other fastenings?

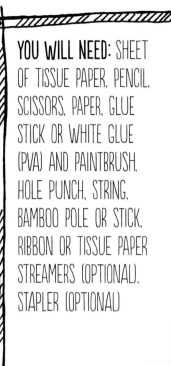 The Sciencey Bit

The air blows through and around the carp kite which creates a **force** called **lift** which, surprise, surprise, helps lift the kite up into the air. If you hold the stick still, you can assess the direction and strength of the wind by seeing where the kite points and how far it lifts up.

YOU WILL NEED:
NETTING (E.G. FROM FRUIT BAGS), STRING OR TWINE, NESTING MATERIALS (E.G. HAIR, YARN, MOSS)

Hang up a bird nesting net

Birds are no slouches. Not only do they get up super early and fly around all day finding food, they even build their own houses. Quite frankly, it makes us look downright lazy. Still, you can make amends by helping them in the soft furnishings department.

Yes, that's right. Even though most nests are made of twigs, birds aren't afraid of a bit of comfort. So why not gather together some suitable materials and hang them up for the birds to find?

If you have an old peanut or suet feeder with gaps in, you could use this, but otherwise it's very easy to make your own bird nest net instead. Take some old netting and make sure the bottom is closed. If there are gaps in the base, weave a piece of string or twine in and out of the netting holes just above it and then pull them tight and tie a knot to seal the gap closed.

You can use lots of different materials to fill the net, such as pet and human hair (that's right, your hairbrush is an interior designer's dream in the world of birds), fur, dried grass and moss, the fluff from a clothes dryer (mmm … delicious!), yarn (cut into 4–8-inch lengths) and feathers.

Put these into the net, leaving some ends of the materials sticking out to encourage the birds to take them. Then seal the top of the net closed as you did the bottom, but with an extra-long piece of string. The ends of this can then be used to attach the nesting net to a tree.

If you haven't caused enough mayhem yet:

Watch to see which materials are most attractive to the birds, and if any **species** are particularly keen on certain types. You can even see if you can spot where the birds are building their nests— just make sure you don't disturb them.

The Sciencey Bit

In watching the birds and how they take and use materials you are practicing **zoology**— the observation and study of animals and their behavior and habitats. In particular, you are practicing **ornithology,** which is the study of birds.

Create newspaper gift bags

YOU WILL NEED: NEWSPAPER, SCISSORS, PENCIL, RULER, CARD, GLUE, HOLE PUNCH, STRING OR RIBBON

Yes, that's right, you can turn old newspaper into free gift bags. This doesn't just save you loads of pocket money but will also make adults believe you are a thoughtful and creative child … rather than a cheapskate. Adults are weird like that.

A single piece of newspaper can only support something light, so begin by placing two sheets on top of each other and cutting out a rectangle (9 inches x 16 inches for small bags or 14½ inches x 24½ inches for large). The extra layer will strengthen your bag so it can hold heavier weights.

Now use a pencil and ruler to mark out the lines shown in the diagram and then fold along each one to form creases.

To help **reinforce** the bag's rim, cut a piece of card the same size as the top flap on each side. Now fold over the flaps, placing the card underneath and stick it down (you will need a second line of glue to stick down the upper sheet).

Add a line of glue on the outside of the tab and bend over the front panel to meet this (you'll need a bit more glue to stick down the extra layer of paper).

Stand your bag up on its end and carefully bend in the bottom flaps so they meet in the middle to make the base. When you do this, you should see that they also meet to form triangular shapes on either side of the bag. Smooth the edges of these between your fingers and thumb to make strong creases so they lie nice and flat. Add glue to the top of your triangular-shaped flaps, stand the bag upright and then press the base of it flat from the inside onto these glued flaps.

Cut a piece of card the size of the base and glue this on the inside to reinforce the bottom of the bag. Next, punch holes in the top rim and thread through string or ribbon of equal length before tying them in a knot to make your handles.

Finally, and most importantly, calculate how much money you have just saved by not buying a gift bag and see how much candy this will buy.

If you haven't caused enough mayhem yet:

Try changing the measurements to make taller, shorter, wider or narrower bags. You can **experiment** with different numbers of layers for bags and see how much weight each design can hold.

 ## The Sciencey Bit

This project helps you understand the properties of everyday materials. Sheets of newspaper have a high amount of *flexibility*—which makes them easy to fold—but not much *strength*. The card has greater strength and also *stiffness*. This makes it the perfect material to **reinforce** the bag at the top and base where it needs it most.

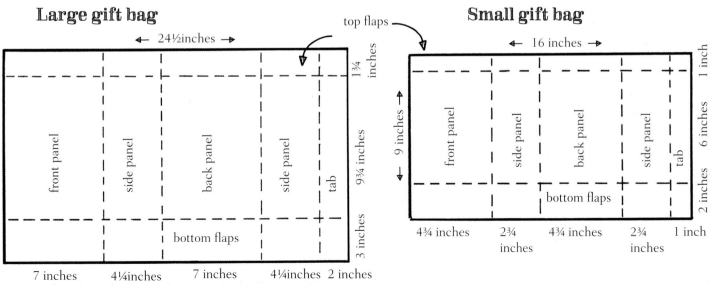

Large gift bag

← 24½inches →

top flaps

	front panel	side panel	back panel	side panel	tab

1¾ inches

9¾ inches

bottom flaps

3 inches

14½ inches

7 inches   4¼inches   7 inches   4¼inches   2 inches

Small gift bag

← 16 inches →

9 inches ↑

	front panel	side panel	back panel	side panel	tab

1 inch

6 inches

bottom flaps

2 inches

4¾ inches   2¾ inches   4¾ inches   2¾ inches   1 inch

Craft a paper towel wreath

What color is a black marker? And yes, this is a trick question because even I don't know the answer. In fact, the only thing I do know is it's not really black (confusing, eh?).

To find out for sure, draw a small circle with your marker in the center of a piece of paper towels—make it nice and thick so there's plenty of ink.

Now fold the paper in half, then quarters, and place it, pointed end down, in a glass with a couple of centimetres of water at the bottom. As the water moves up the paper, it will begin to separate out the ink into different colors so you can see what's *really* in the black.

Do this several times and then leave your paper to dry. Now fold each sheet back into quarters and cut it as shown in the diagram.

Separate the layers of the small center section, fold each into quarters and stick the base of these to the end of a strip of masking tape. Now separate the larger pieces and add these, one after the other, to the masking tape, pleating the paper as you go. Finally, roll up the tape to form a flower.

Make a wreath base by drawing around a large plate on your card and then drawing around a small plate in the center of this first circle before cutting out the ring shape that's formed.

Put a lump of sticky tack underneath the card wreath and push through the end of your scissors to form a slit. Pass the base of your flower through this and then tape it securely to the back of the wreath. Repeat until your wreath is completely covered. Finally, tie a piece of ribbon around the top to hang it from a peg, handle or nail.

And there you have it: a very colorful black marker wreath.

If you haven't caused enough mayhem yet:

Try this using different black pens and also a variety of colored pens. Which color is made up of the largest number of inks? Does this work as well with printer paper, tissue paper or card, and if not, why do you think that is?

The Sciencey Bit

So, how does the addition of water turn your black felt tip into this array of colors? The water is a **solvent** which **dissolves** the ink **pigments** (the **solute**). The pigments that dissolve more easily will travel farther along the paper, which causes the different colors to separate out. This process is called **chromatography**.

Making the paper wreath

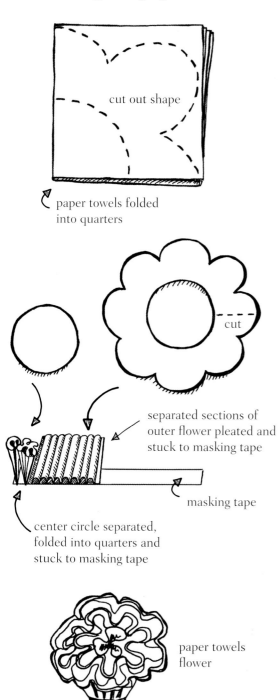

cut out shape

paper towels folded into quarters

cut

center circle separated, folded into quarters and stuck to masking tape

separated sections of outer flower pleated and stuck to masking tape

masking tape

paper towels flower

masking tape rolled tightly

YOU WILL NEED: BLACK MARKER PEN, PAPER TOWELS, GLASS, WATER, SCISSORS, MASKING TAPE, SMALL AND LARGE PLATE, CORRUGATED CARD, STICKY TACK, RIBBON

Make a secret message card

Cards are nice, but let's face it, presents are better. Except for these cards, which are so good you'll be forgiven for not turning up with a gift.

First, take your sheet of card and fold it in half—this will be the base of your card. Now create a template by drawing a simple shape—hearts, stars and circles work well—on another piece of card and carefully cut it out from the center. Do this by folding the card in half and making a small snip in the middle. When you open it up this slit allows you to get your scissors in and cut from the middle outwards.

Lay your template on some card or paper, choosing a color which will contrast well with your base card. Draw around it then cut this out and you are now ready to write your message on the shape, maybe "Thank you!", "Happy birthday" or "This *is* your present!".

Use your template to draw around and cut out the same shape but this time from a material that is going to protect your message but still allow it to be seen: clear contact paper.

Stick your message to the front of your card and then place on the clear contact paper. Now attach the template on top of everything, securing it in place with some paperclips.

You need to hide your message by painting over it—with the template stopping you getting paint anywhere else on the front of the card and the waterproof clear contact paper keeping your message beneath protected.

When the paint is fully dry and the message has disappeared, remove the template and you're free to write inside the card—including a note to tell your friend they must scratch off the front to reveal their secret message.

And if you tape a small coin to the inside of the card your friend will have something to scratch with—and you will prove beyond doubt that you're an *incredibly* generous person.

If you haven't caused enough mayhem yet:

You could **experiment** with different-colored paints to work out which is best at covering the message. Why do you think this is? Can you find another material that would work in place of the clear contact paper?

 ## The Sciencey Bit

This project is all about selecting materials with the right properties. The paint has water in it until it dries, so you need a material that is **waterproof** to protect your message. You also need a material that is **transparent**—which means you can see through it to read your message. It should also be **smooth** so that the paint will easily flake off when scratched with the side of a coin and one side has to be **adhesive** (sticky) so that it attaches to the card.

YOU WILL NEED: SHEET OF COLORED CARD, CARD FOR TEMPLATE, PENCIL, PEN, SCISSORS, DIFFERENT COLORED PAPER OR CARD, CLEAR CONTACT PAPER, PAPERCLIPS, PAINT, PAINTBRUSH, COIN (OPTIONAL), STICKY TAPE (OPTIONAL)

Build a mini greenhouse

Some plants prefer a bit of warmth to get growing, especially if they come from a hot country. So you need to remember this if you're sowing **seeds**.

Hang on—what are you doing? Put the comforter away, and stop filling that hot water bottle now—it's not going to help. Instead, we're going to make our own mini greenhouses.

To do this, take an old, clean plastic bottle, remove the lid and squash it flat. Now use scissors to carefully cut off the top section, press it back into shape and then put the lid back on.

Next, choose a plant pot that is very slightly larger than the bottle greenhouse and fill it with potting soil, firming it down as you go. Sow your seeds according to the packet instructions and give them a good watering.

To make a label, on a small piece of cardboard draw a picture of the plant you have just sown, punch a hole in it, thread through some twine, string or ribbon and tie this around the bottle-top neck. You could also write the name of your plant and the date it was sown on the label.

Finally, move your pot to a sunny windowsill and place your bottle greenhouse over the top.

As the sun warms the air inside the greenhouse it will also cause the water to **evaporate**, but rather than escaping, this will gather on the sides of the bottle as **condensation** and drip back down when the air cools down in the evening. If there is too much condensation, take the lid off the bottle greenhouse for an hour or two to allow some moisture to escape.

When your seedling is nearly touching the top of your greenhouse, make sure you take the lid off or it might damage its growth. But if you're still worried it won't feel at home, you could always put a tube of sunblock and a pair of sunglasses next to the pot—that should do the trick.

If you haven't caused enough mayhem yet:

Try sowing two pots with the same seeds but only add your greenhouse to one of them—what difference does it make to the rate of **germination**, and growth? How about making larger greenhouses out of bigger bottles for growing larger plants?

👓 The Sciencey Bit

The sun's rays can pass through the plastic and will warm the air inside. They also heat up the soil as this is a dark color which **absorbs** more of the sun's **energy**. However, plastic, like glass, is not a good **conductor** of heat (something that allows heat to pass through it) so the air inside remains much warmer than the air outside. The soil and **seed**, and eventually the seedling which emerges, get nice and warm—the perfect conditions for **germination** (starting growth).

YOU WILL NEED: PLASTIC BOTTLE WITH LID, SCISSORS, PLANT POT, POTTING SOIL, SEEDS, WATER, CARDBOARD, PEN, HOLE PUNCH, STRING OR RIBBON

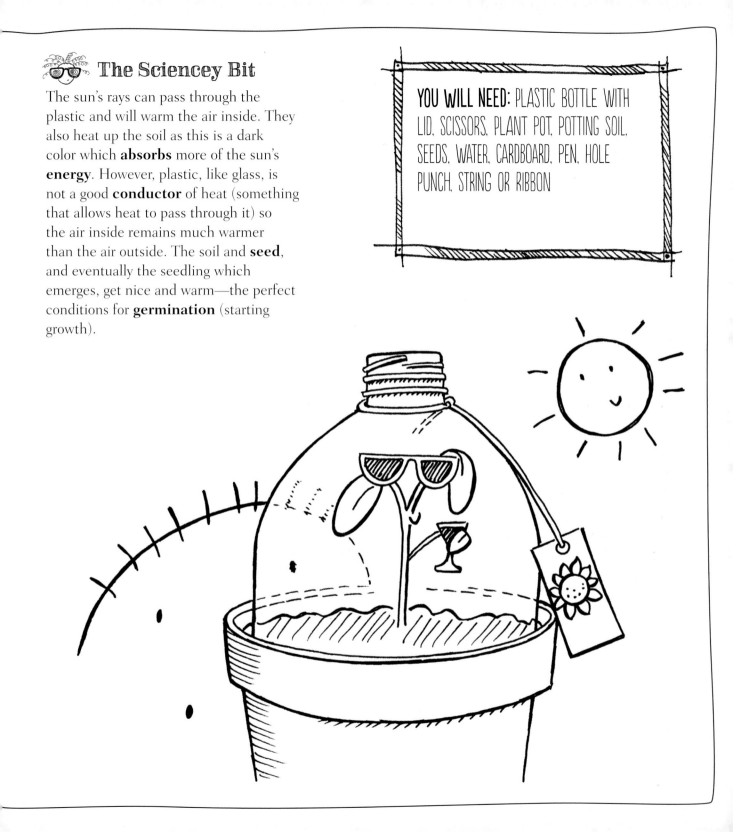

Impale a potato with a straw

There is, of course, no good reason to skewer a potato with a straw—unless you like your chips with holes—but it is a brilliant way of showing off, so who cares?

First, take your raw potato and let your audience examine it so they can see you haven't cheated by baking it first. Now hold it carefully with your fingers and thumb on opposite sides (you don't want to stab these by accident) and pick up a straw in your other hand.

Without making it obvious, put your thumb over the end of the straw and then quickly stab it through the top of your potato so you can see the end sticking out the other side.

You can then let the audience try it themselves. If they attempt this too slowly, the straw will crumple, and if they don't cover the end of the straw they are unlikely to get it all the way through the potato. So, basically you can watch them struggle and flounder for ages. In fact you might even have time to cook those holey chips. Result!

If you haven't caused enough mayhem yet:

There are *so* many other vegetables just waiting to be attacked by a straw! Go on—you know you want to! You could also see how different **diameters** of straws work or find out if chips taste better with holes in them.

The Sciencey Bit

Covering the top of the straw traps the air inside and means the **air molecules** are squashed closer (**compressed**) together as you stab the straw at the potato. The compressed air stops the straw collapsing and makes it strong enough to pierce the potato and go right through. If it's not covered, the air escapes and the straw crumples inwards as you press on it.

Tip: Try to use a good-sized potato as this gives you a larger area to hit with the straw.

Compete in a paper helicopter race

Paper helicopters are easy to make, but a lot more complicated to perfect. Of course, I could tell you the ideal way to make one … but where's the fun in that?

Instead you can use this basic design, cut from a piece of paper (see diagram) to try and work out the very best combination for the longest flight. You could do this simply as an **experiment** on your own. But if there are a few of you, why not set the timer and give yourself a half hour to come up with your ultimate design?

To prevent industrial espionage, it might be worth working on your designs and test flights in separate rooms. And just remember, adult intervention is not allowed (and let's face it, this might not help anyway).

The main elements you can alter in creating the ultimate flying machine are body length, body width, rotor length, rotor width, amount of paperclips added, bends added to your rotors and the thickness of paper you use. It helps to note all these factors in a table and record the test flight times for each factor you change.

When the time is up, you could let everyone have three flights. The helicopters must be dropped from the same height and someone needs to record or count each attempt's length of flight—with the longest being declared the ultimate winner (and having their design copied by everyone else).

If you haven't caused enough mayhem yet:

You could add in some more design factors to alter and test—how about fixing tape to the blades or body of the helicopter? You can also add decorations and even a pilot to your finished helicopter. This probably won't help you win, but it'll look good—which is nearly as important.

The Sciencey Bit

I don't want to shock you but you have been working scientifically without even knowing it! You've run **comparative and fair tests**, collected and analysed **data**, and used this to select materials and adjust your design. And you thought you were just trying to beat your friends.

Making the helicopter

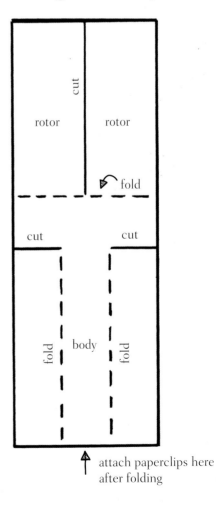

rotor	cut	rotor

↪ fold

cut   cut

fold   body   fold

↑ attach paperclips here
after folding

Blow a square bubble

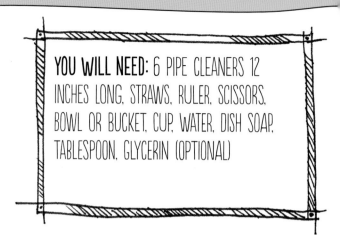

Round bubbles? You blow round bubbles? Oh that's SO last year. This season it's all about the square bubble, so pay attention.

To build your square bubble maker you'll need to cut your pipe cleaners in half so they are 6 inches long and then cut twelve 3½-inch lengths of straw.

Now twist the ends of two pipe cleaners together and slip on two pieces of straw to make a right angle. Twist another pipe cleaner on each end and add another straw to each of these and then finally twist the two spare ends together to form a square.

Now add more pipe cleaners and straws to each corner pointing upwards, and then finally, attach the last four pipe cleaners and straws to form a cube.

Make your bubble mixture in a large bowl or bucket by adding three cups of water to a cup of dish soap and mixing together slowly so it doesn't froth up. If you have it, you can also add a tablespoon of glycerin to the mixture for longer-lasting bubbles.

Finally, dip your bubble maker into the mixture so it's all submerged (you can squish the cube a little to make this easier and reshape it afterwards) and slowly bring it out. There should be films of bubble mixture on all edges, meeting in the middle. Now dip a straw into the bubble mixture and … NO, don't suck … it really won't taste very good! Instead, place the end of the straw into the center of your bubble maker and gently *blow*. Amazingly, you should see a square bubble start to form at the center.

Although, square bubbles are SO five minutes ago. The fashion right now is for triangular bubbles—hadn't you heard?

If you haven't caused enough mayhem yet:

Try building different shapes with your straws and pipe cleaners. What happens when you dip this into the bubble mix?

 # The Sciencey Bit

Water has a skin because water **molecules** are **attracted** to each other and stick together in a layer. This is called **surface tension**. Soap mixed in with the water makes this surface layer or "skin" stretchier but it's still not very strong, so bubbles always use the smallest **surface area** to enclose the air inside. This is usually a sphere but here the soap film is stretched between all the side struts causing the bubble at the center to appear square (although, if you look carefully, the sides still bulge).

Send timed-release water lily messages

Yes that's right—forget phone calls, texts or emails—you are going to communicate via water lilies, and that's *much* more impressive.

First, you need to make your water lilies. You can trace the outline of the design opposite (see diagram) and then transfer this onto card or else recreate it using pencil, ruler and your own artistic brilliance.

Cut out the water lily shape from the card to form a template. You can now draw around this onto a selection of different types of paper or card. Take one water lily from each and fold in all the petals to the center, one at a time, in a clockwise pattern. Next, place the folded flowers in a basin of water and start the timer, making a note of which lily opens at which time.

Now you have the timing **data**, you can make your message. Cut out a circle or even a petal pattern that will fit in the center of a water lily, stick this on with a glue stick and write your message using pencil, or wax crayon (as these won't **dissolve** in water). And to really impress people, you can write a separate part of your message on different types of paper or card with the first word or sentence on the fastest opening lily and gradually working towards the slowest.

You could use this to spell out someone's name a letter at a time, write "Happy" "Birthday" "to" "you", "Be" "my" "Valentine", or even "Yes" "I" "know" "texting" "would" "have" "been" "faster".

Water lily template

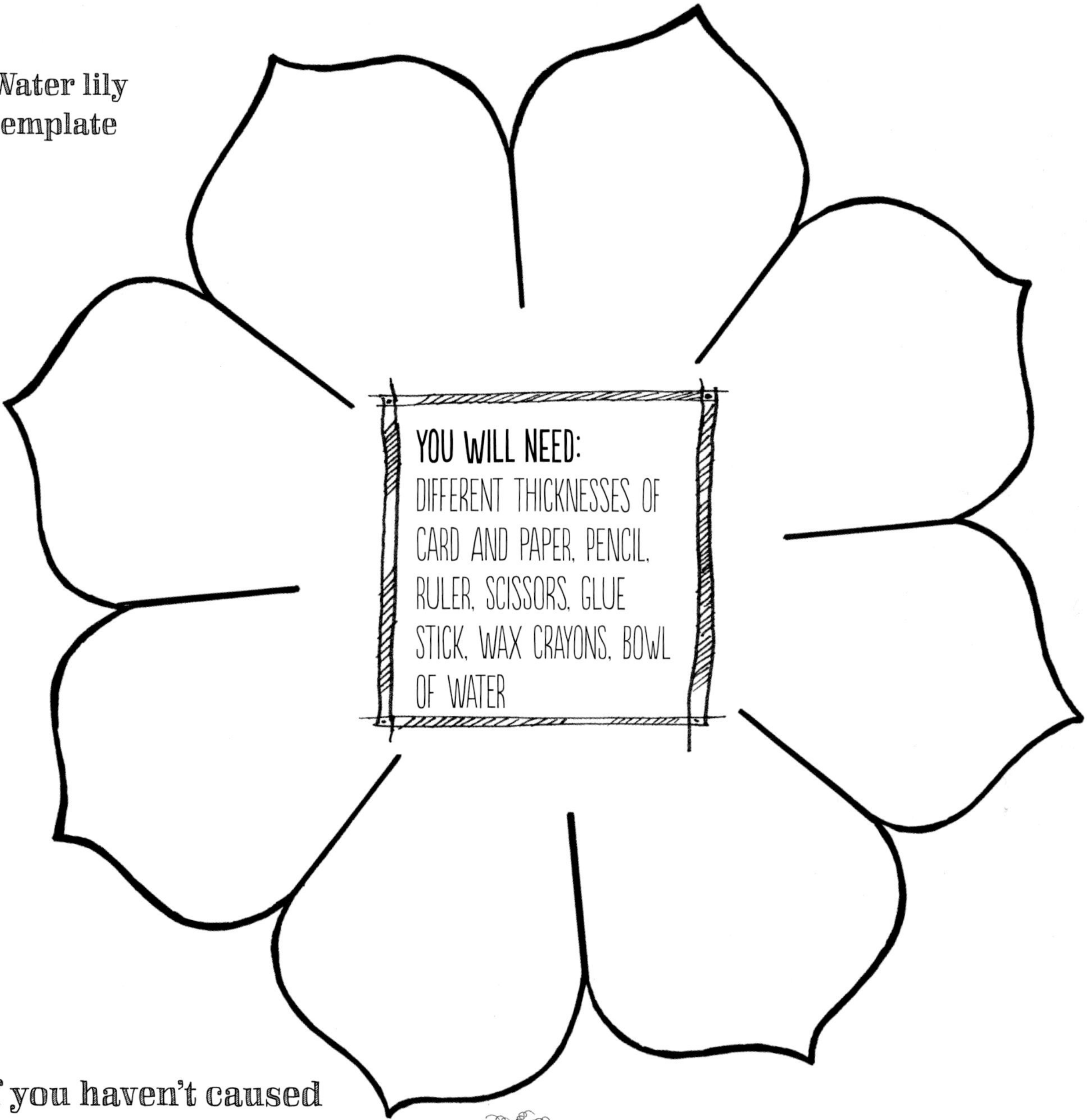

YOU WILL NEED:
DIFFERENT THICKNESSES OF CARD AND PAPER, PENCIL, RULER, SCISSORS, GLUE STICK, WAX CRAYONS, BOWL OF WATER

If you haven't caused enough mayhem yet:

Keep **experimenting** with different types of paper. What happens if you use newspaper or paper towels? What about baking parchment or tracing paper?

The Sciencey Bit

Paper and card are made of tiny wood **fibers**. When these **absorb** water they swell up and expand and this causes a little movement—enough to make the flower open. Different types of paper absorb water at different speeds depending on how thick they are or if the paper is coated.

Making the bathtub paddleboat

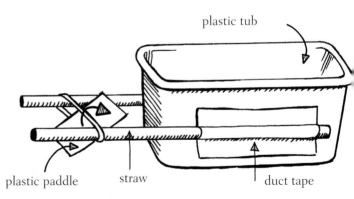

plastic tub

plastic paddle   straw   duct tape

Make a bathtub paddleboat

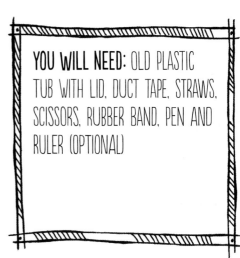

YOU WILL NEED: OLD PLASTIC TUB WITH LID, DUCT TAPE, STRAWS, SCISSORS, RUBBER BAND, PEN AND RULER (OPTIONAL)

If you have to wash, then at least you can add a little excitement to bathtime by creating your own mini paddleboats. It might even distract you from the trauma of getting clean.

First, take the empty plastic tub and use duct tape to connect a straw to either side so that the ends of these stick out from the back (see diagram).

Now use scissors to cut a square shape from the plastic tub lid—you can do this by sight or use a ruler and pen to be more accurate. Just make sure it's narrower than the space between the two straws.

Finally, hook a rubber band over the end of the straws, slip your square of plastic in between and start to turn it over again and again. You should see the rubber band twisting at the sides.

When you have wound it plenty of times, place the boat in your bath and let go. As the elastic unwinds, it will cause the plastic square to move, pushing the water away and propelling the boat in the other direction. In fact, it will look very much like it is desperately trying to escape from the bath. I'm sure you know how it feels.

If you haven't caused enough mayhem yet:

Try winding the square in different directions— what effect does this have? What happens if you add two more straws and a second paddle to the other end of the boat? Try adding weight to the boat—maybe something scary like a bar of soap. What happens then?

The Sciencey Bit

As you twist the rubber band you are changing movement energy (called **kinetic energy**) into stored up energy (called **potential energy**). When you let go, this potential energy is changed back into kinetic energy as the band unwinds, making the paddle move.

Create a snowstorm

No we're not going to create a real snowstorm. I've told you before, you are NOT a weather god or superhero and you cannot magic up snow days however hard you try.

So now we've scaled back your ambition a little, let's create a snowstorm … in a jar.

First, you'll need a small, waterproof object. Plastic figures work well but just make sure they're big enough to be seen easily in the jar and they aren't your sister's most precious ornament. Now take your waterproof glue and attach your figure to the inside of the jar lid and leave it to dry. Yes, sorry, you'll have to be a little patient now—which is probably harder than conjuring up snow.

Next, add your glitter to the jar, then fill it nearly to the top with water and pop in a few drops of glycerin. Finally, screw on the lid tightly, give it a shake and turn it upside down. Now you can watch as your glitter whirls in a snowstorm **suspension** before finally settling to reveal … your sister's most precious ornament! Ooooh, you're in trouble now!

If you haven't caused enough mayhem yet:

Try adding different amounts of glycerin to the water and time how long each mixture makes the glitter swirl. Or you could just time how long it takes your sister to start screaming after she spots what you've put in your snowglobe.

 The Sciencey Bit

As you shake it up, the fine glitter forms a **suspension** in the water that will eventually settle out. Glycerin is a thick **liquid** that **dissolves** in water so the more you add, the greater the water's **viscosity** (its thickness) which means the longer it takes the glitter to settle.

Tip: Draw a sun near the top of the card and rain near the base to help you read the barometer's predictions.

YOU WILL NEED:
LARGE BALLOON, SCISSORS, WIDE-NECKED JAR, RUBBER BAND, STRAW, STICKY TAPE, CARD, RULER, PEN

Build a barometer

A **barometer** can help you predict the future but sadly, only in terms of the weather. For future predictions of how well you will do in your math quiz, you will have to turn to the foolproof methods of crystal balls and horoscopes.*

First, cut off the lower third of your balloon and then stretch this over the top of your jar. Make sure there is no dimple in the middle—it must be stretched nice and flat—and then secure it in place with a rubber band.

Cut one end of a straw at an angle so it "points" and tape the other end to the middle of your balloon-covered jar top.

Bend your card in half so it stands up and position it behind the pointed straw. Make a mark where the straw currently points and then take your card and carefully add more marks above and below at ¼-inch intervals before placing the jar and card somewhere they won't be disturbed (on top of your math books, for example).

Keep checking on your barometer at regular intervals throughout the day. If the straw moves upwards this is signalling drier weather on the way, if it moves downwards it indicates that it is more likely to rain. And if it stays where it is, nothing changes (including your chance of doing well in that math quiz unless you revise at some point).

*Except they don't work. Sorry. You'll just have to study. Strangely enough that tends to help.

If you haven't caused enough mayhem yet:

Try using an extra-long straw to see what difference this makes. To do this, take a pair of straws and cut two slits about ¼ inch long at the end of one straw before pushing this into the end of the other straw and taping them together.

The Sciencey Bit

The jar is full of **air molecules** trapped there by the balloon while the air outside is free to move. As cool air sinks it puts more pressure on the jar, meaning the balloon flattens or dips causing the straw to point upwards.

As warm air expands and rises it lowers the **air pressure** outside the jar but the air inside pushes outwards by the same amount making the balloon dome and the straw point down. This helps predict weather because low air pressure often leads to unsettled and wet conditions.

Make wax resist leaf pictures

Some things don't get on—yep, not even when an adult says "Play nicely, you two!" in threatening tones. Take wax and water for example—they can't stand each other. But unlike you and that kid next door, this isn't personal, it's just science.

And the fact that these things repel each other can be very useful—at least for painting some impressive pictures.

To create your artwork, place a leaf under your paper and then rub over this area with the side of your candle, pressing down hard. Keep doing this again and again, moving the leaf underneath different areas of your paper as you go.

Now you need to paint over the top. You can buy special watercolor paints for this, but it's cheaper and easier to mix some ordinary poster paint with water instead. Use a brush to apply a thick layer of watery paint over the top and you should begin to see the leaf patterns appearing through it. This is because the wax, which rubbed off over the raised veins and edges of the leaf, repels the water in the paint.

When you're happy with the effect, leave the paint to dry before framing it … and showing it to the kid next door with the words "See, this paint finds wax as repellent as I find YOU!" and then probably running away … FAST!

If you haven't caused enough mayhem yet:

Why not find other textured surfaces or objects to put under your paper and see what sort of patterns they create? Or try this project with white or colored wax crayons instead—or cardboard instead of paper—what difference do these things make?

The Sciencey Bit

Watery paint runs off the candle-rubbed area because wax is **hydrophobic**—this means it repels water. The paper is mostly made of **cellulose** which is **hydrophilic**—this means it **attracts** water so it will quickly soak up the watery paint.

YOU WILL NEED: LEAF,
PAPER, WHITE CANDLE, PAINT,
PAINTBRUSH, WATER

Tip: Use a variety of paints to make a
really colorful wax resist picture.

Bring ghosts to life

That's right, we are going to bring the dead back to life! Cue the mad scientist laugh: MWA HA HA!*

First, you're going to need a few ghosts. If you don't actually live in a haunted mansion it's best to make your own. Do this by cutting out a ghostly shape from a piece of tissue paper and then use your black marker pen to add spooky eyes and a mouth. If you want to make several, you can lay sheets of tissue paper on top of each other and then cut them out all in one go.

Next, use a piece of sticky tape to attach the base of the ghost to a table so it can't fly away (they can be tricky like that, these supernatural beings).

Blow up a balloon and tie the end in a knot before rubbing it over your hair or a sweater for a minute. You could also use a plastic ruler or plastic comb for this. Yes, you know, a comb. No, it is not a weapon of torture—it's just something you use to make your hair look tidy. What do you mean "That's the same thing!"?

Finally, take your balloon, ruler or comb (stop looking so scared), hold it above the shapes, then move it nearer and nearer and watch as the ghosts rise up!

For added atmosphere you could make a few tombstones from card, or even a haunted mansion, turn the lights low, and play some spooky music. Then again, you could just show people the comb … and then tell them what it's for. That should cause a few screams.

*Actually, we're really just going to make a bit of tissue paper move—but that doesn't sound scary so let's just be quiet about this and keep cackling.

If you haven't caused enough mayhem yet:

Try making ghosts out of different types of paper—which works best? Does it make a difference if you rub your hair, or the jumper, for longer? How long does it actually take you to comb your hair—and can anyone recognize you afterwards?

The Sciencey Bit

When you rub the balloon against your hair or jumper, invisible **electrons** (tiny **particles** with a negative charge) build up on the surface of the balloon. The electrons have the power to pull very light objects (with a **positive charge**) towards them—in this case, the tissue ghost.

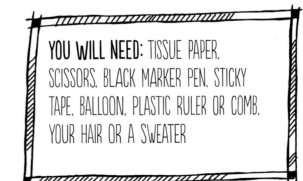

YOU WILL NEED: TISSUE PAPER, SCISSORS, BLACK MARKER PEN, STICKY TAPE, BALLOON, PLASTIC RULER OR COMB, YOUR HAIR OR A SWEATER

Make seed cards

Seeds are very good at sleeping—they're even better at it than you are on a school day morning. In fact, they won't wake up when you pull back the covers, shout "YOU'RE LATE!"—or stick them to the front of a card. Which is very handy because we're going to do just that.

Start by drawing a simple bold shape on some pale colored card, then cut it out carefully. Now fold a darker colored sheet of card in half before sticking your pale shape to the front.

Next write your message inside and don't forget to include instructions such as "You can grow this card. No really—you can! Just lay it flat, spray or soak it with a little water, keep it damp and in a few days you will have some tiny leaves to taste."

In the bowl mix together one tablespoon of flour and one tablespoon of sugar with a little water until it forms a smooth paste. Next, paperclip the card together so it lies flat and use a paintbrush to create patterns with the paste on your design. You can then sprinkle on your seeds, like glitter, so they stick to the paste.

When the card is dry, shake off any excess seeds, take off the paperclips and carefully slip the card into the envelope. And don't worry—you won't have to whisper while you do it—the only thing that will wake these seeds is throwing a bit of water at them. In fact, I wonder if that would work for you on a school day morning? Maybe we should give it a go.

If you haven't caused enough mayhem yet:

Experiment by using different seeds on the cards—how long does each take to start growing? Does it make a difference if you sprout the card in the dark or light? What about if you add juice instead of water? Or cover the seeds completely in water? Or place them in the refrigerator?

The Sciencey Bit

A **seed** is very dry and needs water to begin growing (**germination**). The paste dries quickly so the seed can't **absorb** enough water from it to start growing. However, when you keep the card damp for longer, the seed takes in more water and it starts to grow.

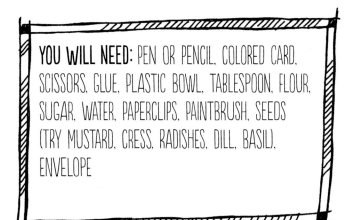

YOU WILL NEED: PEN OR PENCIL, COLORED CARD, SCISSORS, GLUE, PLASTIC BOWL, TABLESPOON, FLOUR, SUGAR, WATER, PAPERCLIPS, PAINTBRUSH, SEEDS (TRY MUSTARD, CRESS, RADISHES, DILL, BASIL), ENVELOPE

Build a sandcastle city

So you're at the beach with bucket and spade. What do you do? And just remember, if you say "Build a sandcastle," I shall be *very* disappointed. After all, you have vision, you have ambition, and you have a lot of sand at your disposal.

A sandcastle city? That's more like it. Now you're thinking big!

First, you need to find the ideal spot. Go for somewhere below the shoreline so the sand will be damp when you dig into it—after all, you need access to the perfect building materials.

Now set out your city perimeters and remember to make these nice and large. If there are lots of you helping out, make it bigger still so there is space to work without you accidentally destroying each other's genius creations and starting a sandcastle city civil war.

If you want to give your citizens some added protection it's a good idea to dig in a large moat around the edge. What's more, the damp excavated sand is very useful to form hills or city walls inside the moat.

Remember to be as creative and artistic as you like. You could conjure up some craggy cliffs

(wet sand trickled from your hand is perfect for adding this effect) or find different shells, pebbles, or seaweed from the beach to add landscape details, roads, and decoration. And yes, you can even add some sandcastles too.

If you haven't caused enough mayhem yet:

If you're relatively near the sea itself you could try to dig a channel to the water's edge. This way, if your moat and channel are nice and deep, they will start filling up as the tide comes in. Or if the water level is quite high, moats and lakes might even begin to fill naturally as you dig down.

The Sciencey Bit

Dry grains of sand easily move past each other but when you add water it surrounds each grain and the **surface tension** of the water makes the grains of sand stick to each other, which helps create structures like sandcastles. If you add too much water the sand will become **saturated** (this means it contains as much water as it can hold). Adding any more water than this makes the grains of sand separate from each other and your sandcastles collapse.

YOU WILL NEED:
BUCKETS, SPADES, SAND, AMBITION

Freeze an ice bowl

Ice is amazing at preserving things. And no, I'm not just talking about ice cream. When something is completely surrounded by **ice** it stops it from **decaying**—whether that's leaves, fruit, flowers, or woolly mammoths. Which is good because these also look gorgeous in an ice bowl (well, maybe not the woolly mammoth).

Begin by adding about 2 inches of water to the bottom of the larger bowl and place it in the freezer for a few hours. If there is not enough room, you may need to eat some ice cream … just to clear space, of course.

When the water has frozen, take the bowl out and start wetting your selected flowers or leaves before sticking them to the inside of the bowl. Next, add your second bowl inside the first, weighting it down with some large stones.

Now add four lumps of sticky tack, evenly spaced and molded around the top edge of the outer bowl to stop the two bowls touching each other. Take your two lengths of crêpe bandage or old pantyhose –place one under the outer bowl and then bring the ends back up top to tie tightly in a knot. Do the same with the other piece but lay it at right angles to the first—this will help hold the bowls in place (see diagram).

Finally, pour water into the gap between the bowls until it's nearly at the top and then put your soon-to-be-ice bowl back in the freezer overnight.

Next day you can take it out and leave it on the side for a few minutes until you are able to loosen the bowls. Now carefully untie the bandages, remove the stones and take off the inner and outer bowls to reveal a beautiful ice bowl—perfect for displaying fruit … or lots and lots of ice cream.

You can store your ice bowl and its preserved flowers or leaves for as long as you need—just wrap it in foil and place in the freezer. What's that? You need to eat more ice cream to make room for it? Oh, okay then!

If you haven't caused enough mayhem yet:

Try making smaller or larger ice bowls—which ones melt the fastest? What about ones with thicker or thinner bases or sides?

The Sciencey Bit

When things **decay** it is because **bacteria** (tiny living things, too small to see without a **microscope**) are breaking them down to use as food. When things are encased in **ice** they are preserved because bacteria cannot do their job in very cold temperatures. This stops the flowers from decaying so they remain looking fresh and lovely.

YOU WILL NEED: TWO BOWLS (ONE LARGE, ONE SMALLER), WATER, FLOWERS AND/OR LEAVES, STONES (OR OTHER WEIGHTS), STICKY TACK, CRÊPE BANDAGE OR OLD PANTYHOSE, FREEZER

Making the ice bowl

crêpe bandage or pantyhose tied around bowls

sticky tack molded over edge of outer bowl

inner bowl

rocks to weight down inner bowl

outer bowl

water (to be frozen)

flowers and/or leaves

ice

Whisk up a lemon mousse

Yes, you did read that right. *Lemon mousse.*

What do you mean "Where's the chocolate?" You can make mousses that aren't chocolate flavored, you know. Yes, really, you can!

First, separate the egg by knocking it on the edge of one of the bowls until a crack starts to form. Now gently push your thumb into the crack to make the egg break open in the middle and then pour the yolk from one half to the other. As you do this, the egg white should fall away and be caught in the bowl below. Once it's all separated, place the egg yolk in another bowl with the sugar.

Whisk the egg white and after a few minutes you will see an amazing change—the small amount of **transparent liquid** egg white has become large, fluffy, white and **opaque**. Ta da!

In yet another bowl, add the cream and start whisking until it also changes from a sloshy liquid to a thick whipped cream. Ta da!

You can stop ta da-ing now—we've more work to do.

Place your gelatin sheet in the flat-bottomed bowl, cover it with cold water until soft then take it out, squeeze it and throw away the water before putting it back in the bowl. Next, ask an adult to pour in some very hot water so it just covers the sheet and then gently stir this until the gelatin **dissolves**. Add this to the egg yolk and sugar and beat them all together.

Carefully grate the rind (the outside of the lemon) then squeeze the juice out of the lemon before adding both to your yolk mix and beating it in. Finally, fold together the egg white, whipped cream and yolk mixture until they are evenly mixed. Do this by using the side edge of a large spoon or spatula to gently lift the mixture up and over itself, again and again. Now pour the mousse into a serving dish, cover with plastic wrap, and leave it to set in the refrigerator for an hour or two.

When you're ready to serve you could add some decoration to the top. Yes, you're right, chocolate sprinkles would be *perfect* for this.

If you haven't caused enough mayhem yet:

Oh go on then, you can make a chocolate mousse, too! Because it's sweet you can leave out the sugar and as it's not a **liquid** like lemon juice, you don't need the gelatin to make it set. Ask an adult to help you melt 1¾oz of cooking chocolate, then beat in the egg yolk and finally fold this mixture into the beaten egg white, cover with plastic wrap and leave it to set in the refrigerator.

YOU WILL NEED: 1 EGG, 2¼OZ SUPERFINE SUGAR, 4 BOWLS (ONE WITH A FLAT BOTTOM), WHISK, 2½FL OZ HEAVY CREAM, 1 SHEET OF GELATIN, WATER (HOT AND COLD), 1 LEMON, GRATER, SPOON OR SPATULA, SERVING BOWL, PLASTIC WRAP, DECORATIONS (OPTIONAL)

 ## The Sciencey Bit

Whisking produces air bubbles. In the cream these bubbles become trapped by fat **molecules,** making the cream much thicker. The action of whisking also unravels the long **protein** molecules in egg whites so they surround and trap the air bubbles. This changes the **liquid** egg white into a thick white foam.

Tip: Instead of making one large lemon mousse, you can separate it into smaller, individual bowls and serve these.

Making the spectroscope

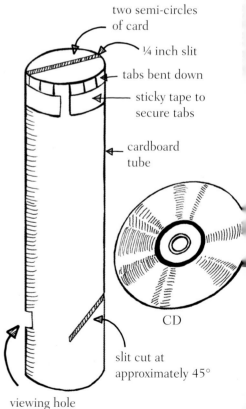

two semi-circles of card

¼ inch slit

tabs bent down

sticky tape to secure tabs

cardboard tube

CD

slit cut at approximately 45°

viewing hole

See rainbows

Believe it or not, the inside of a cardboard tube is the perfect place to spot a rainbow. Still not convinced? Well, grab an old CD or DVD that nobody needs any more and we'll prove it.

First, take your cardboard tube and squash it flat enough so you can cut a slit near the base that goes about halfway through the tube. This should be roughly at a 45-degree angle (you can use a protractor to check). You also need to cut a viewing point about ½-inch square on the opposite side to the slit before squeezing the tube back into shape (see diagram).

Trace around a circle a little larger than your tube (such as a small cup or cup) on a piece of black card and cut this out. Fold it in half and then cut along this line so you have two semi-circles. Now trace over half of the end of your tube on each one and then cut lines between the two semi-circles to form tabs. Place both pieces of card over the top of your tube so there is a narrow slit about ¼ inch wide between the two, then bend down your tabs and tape them in place (see diagram).

Slot the old CD, shiny side up, into the 45-degree slot you cut earlier, and use two long sausage shapes of sticky tack along the top and bottom edges to make sure no light gets through.

Now take your tube outside and as the light enters the slot from above, you can look through the viewing hole and see a **spectrum** of rainbow colors on the CD surface. Yes, that's right, you have a portable rainbow spotter, or more accurately, a **spectroscope**. Now that's the way to upgrade a boring cardboard tube!

If you haven't caused enough mayhem yet:

Try looking at other sources of light through the spectroscope—different types of bulbs or even firelight (with an adult's help)—how do they differ?

The Sciencey Bit

A **spectroscope** is an instrument used to split light into its different **wavelengths**, which we see as the different colors of the rainbow. The surface of the CD is covered in microscopic ridges that **refract** light—or bend it—and this separates the colors. And because the CD's surface is mirrored, this light is then **reflected** back to your eye for you to see.

Create bubble art

Yes, you can blow loads of bubbles with a straw and for once, no adult's going to tell you off!

Oh, and you'll also make some great artwork and learn a bit about science. But really it's ALL ABOUT BLOWING BUBBLES!

First, you'll need to protect your table with a wipeable cloth or old newspaper. Yes, these are messy bubbles. It just gets better, doesn't it?

Now make up your bubble mix. In a small bowl or mug, add two tablespoons of paint to about a tablespoon of water and two big squirts of dish soap. Mix this together really well with a spoon and then put in the end of your straw. You need to blow, quite gently to begin with, until bubbles begin to form. If they're too watery, add a bit more paint. If they burst too quickly, try another squirt of dish soap, and if there aren't enough of them, put in a little more water.

When the bubbles have reached above the top of your bowl or cup, you can make a print of them by placing a piece of paper or card over the top, then take it off and let it dry. Use the prints as an artwork in themselves, for wrapping paper, or gift tags, as a background to another picture, or cut out and use smaller bubble sections to form flowers, or hair, or animals, or even bubbles (although that does show a serious lack of imagination).

If you haven't caused enough mayhem yet:

Try using different sizes and shapes of bowls or trays to blow your bubble mix in—which works best? Does it make any difference if you use wider or narrower straws? Try to calculate your own perfect recipe for a bubble paint mix. If you blow bubbles in a glass of milk, how many seconds will it take before an adult tells you to stop?

The Sciencey Bit

When you add air through the straw it becomes wrapped in soap film, which forms bubbles. The outside and inside surfaces of the bubble are made of a layer of soap **molecules** with a thin layer of water sandwiched between. As there are paint molecules mixed with the water, these will leave a mark on the paper when the bubbles burst.

Tip: Just make sure you don't suck!

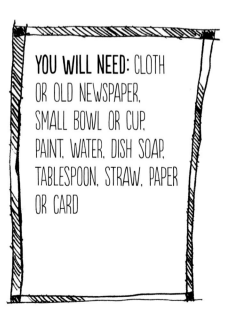

YOU WILL NEED: CLOTH OR OLD NEWSPAPER, SMALL BOWL OR CUP, PAINT, WATER, DISH SOAP, TABLESPOON, STRAW, PAPER OR CARD

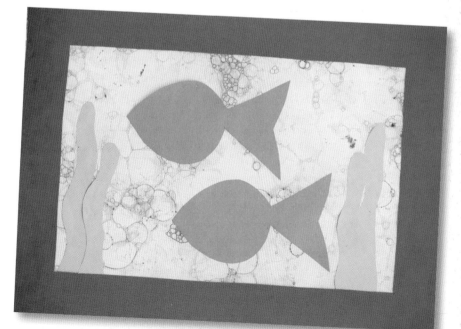

Build a constellation nightlight

Falling asleep under the stars is a nice idea, but probably a bit cold. So why not stay in your nice cozy bed and bring the stars inside with you instead?

Of course, I am not suggesting you transport giant luminous balls of **gas** into the house. That would be silly. But you can recreate some of your favorite constellations in a jar.

First, take your card, roll it in a tube and slot it into your jar. Mark with a pencil where this reaches the top of the jar and where it overlaps with itself. Take it out and draw these lines with a ruler. Now cut the top so it fits the jar and cut the width so it's about ½–1 inch wider than the point where it overlaps. Finally, pop this back in the jar to double-check it fits.

Find a map of constellations (groups of stars) or use some on the opposite page and mark them with a pencil on your tracing paper. Turn this onto the card and transfer the marks by rubbing your pencil on the other side.

When you're happy with your constellation patterns, create holes where each star appears, by pressing the end of a pen or pencil through the card into a lump of sticky tack.

When you've completed your star map, roll it up and put it back into the jar (see diagram). Finally, place a small flashlight or battery–powered tea light inside, replace the lid and turn off the lights. Magically, the constellations will shine out around your room so you can feel astronomically educated—but also snug and warm.

If you haven't caused enough mayhem yet:

Try making different star maps for the various constellations. You could use a larger nib to make the constellations and then smaller pinprick dots for other stars so it looks more like the night sky.

The Sciencey Bit

Each star you see in the sky is a luminous ball of burning **gas**. The nearest star to Earth is our sun. You can look for the constellations—they appear in different parts of the night sky as the year moves on because the Earth is rotating around the sun.

Some popular constellations

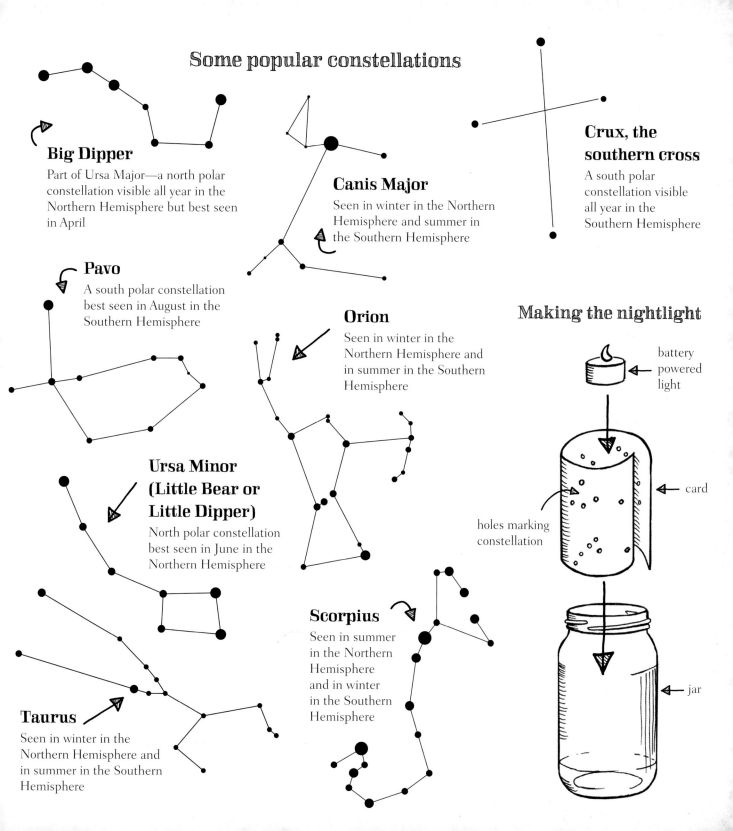

Big Dipper

Part of Ursa Major—a north polar constellation visible all year in the Northern Hemisphere but best seen in April

Canis Major

Seen in winter in the Northern Hemisphere and summer in the Southern Hemisphere

Crux, the southern cross

A south polar constellation visible all year in the Southern Hemisphere

Pavo

A south polar constellation best seen in August in the Southern Hemisphere

Orion

Seen in winter in the Northern Hemisphere and in summer in the Southern Hemisphere

Ursa Minor (Little Bear or Little Dipper)

North polar constellation best seen in June in the Northern Hemisphere

Scorpius

Seen in summer in the Northern Hemisphere and in winter in the Southern Hemisphere

Taurus

Seen in winter in the Northern Hemisphere and in summer in the Southern Hemisphere

Making the nightlight

battery powered light

card

holes marking constellation

jar

Make a gel-tastic air freshener

Let's face it, there are many times when you could do with an air freshener—after your dad's been in the bathroom for half an hour, when the dog's rolled in something, and pretty much any time your brother enters the room.

Thankfully, it's very easy to create your own—and what's more it'll outlast these awful stinks (okay, maybe not your brother's—that'll be around for years).

First, ask an adult to pour 5fl oz of boiling water into the jug. Meanwhile, put the gelatin sheets into a bowl of cool water until they are soft, then squeeze them out and carefully put them into the boiled water, stirring until they have **dissolved**.

Top the jug up with 5fl oz of cold water before adding 10–15 drops of essential oil. Mix in a few drops of food coloring until you're happy with the shade and finally, stir in a tablespoon of salt to stop any mold growing—after all, a moldy air freshener is nearly as grim as a smelly brother.

Now pour your mixture into small jars and leave to set. Finally, place your gel air fresheners around the house—or just next to wherever your brother is standing.

If you haven't caused enough mayhem yet:

Try using more or less gelatin—what effect does this have? Does the scent last any longer? Concoct different fresheners using different oils—which one does the best job at masking your brother's stink?

The Sciencey Bit

These gel air fresheners can keep scenting the air for a long time because gelatin is a **polymer**—which means it's made up of long chains of **molecules**, and these chains weave together trapping fragrance oil **particles** inside. As the gel **evaporates** it frees these trapped particles which causes a continuous scent to be released.

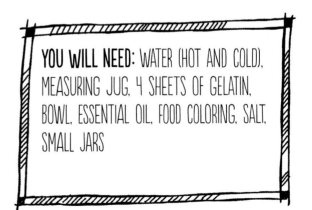

YOU WILL NEED: WATER (HOT AND COLD), MEASURING JUG, 4 SHEETS OF GELATIN, BOWL, ESSENTIAL OIL, FOOD COLORING, SALT, SMALL JARS

Grow pea shoots

Let's make this clear—we're talking pea shoots here, NOT peashooters. The first are delicious and quick-to-grow vegetables, the second are small weapons used to shoot dried peas at unsuspecting victims. As you can see, it's best not to get them muddled up.

Begin by stuffing your jar or plastic cup with the paper towels until it feels full. Now pour in a little water and wait for it be soaked up until they are wet all over. If you can still see some dry pieces after a few seconds, pour in a little more water.

Now take a dried pea (one you haven't already launched from a peashooter) and slot it about ½–1 inch down the side of your jar or cup so it can be seen from the outside. Do this every inch until you have gone all the way around the jar. Finally, place this on a sunny windowsill and let the water do its work.

Over the next few days, the pea will be taking in water from the wet paper towels and swelling until eventually a root emerges from the base, followed a little later by a green shoot from the top. Just make sure you keep the water topped up so the paper towels and peas don't dry out.

When the shoot is about 6–8 inches tall you can harvest your crop. Simply take some clean craft scissors and cut the shoot about 1–1½ inches above the rim of the cup. You can then eat this raw—on its own or in a salad—or add it to stir-fries. What's more, using dried peas to grow a delicious ingredient will go down a lot better than shooting them at your little sister.

If you haven't caused enough mayhem yet:

Try adding a dried pea at a time—one a day—and see if you can chart the entire **germination** process in a single jar. What happens if you place the jar in a cupboard rather than a sunny windowsill? How about if you try this without the water?

The Sciencey Bit

The dried pea is the **seed** of the pea plant. It needs water to start it into growth—or **germination**. When it soaks water up from the napkin through a tiny hole (the **micropyle**) in its seed coat (the **testa**) this enables it to grow its first root (the **radicle**) and first shoot (the **plumule**).

Build a sand clock timer

Parents have an odd concept of time. If you ask them when they'll be ready to drive you to your friend's house, make that hot chocolate they promised, or help with your homework, they'll almost always say "In a couple of minutes". But it is never just a couple of minutes— and now you can prove it.

To build your "parent-check" sand clock, first take two small plastic drinks bottles of exactly the same size. Make sure they are clean and leave them to thoroughly dry—this is important as, if any moisture is left inside, your clock won't work properly.

Now remove the lids and trace around the top of the bottle onto a piece of card before cutting this out carefully. Place it on a piece of sticky tack and use a sharp pencil to make a hole in the middle. The hole will be bigger or smaller depending on how far you twist through the nib—but it's best to start small (you can always make it bigger later). Then use a couple of pieces of masking tape to attach the disk of card to the top of one bottle.

Measure out a set amount of sand, say 2fl oz, and use a funnel to add this to the open-ended bottle. Then lay both bottles on their side so the ends are pushed together and tape them together with more masking tape.

Set a stopwatch or use a clock with a second hand to time as you tip the bottles upright, with all the sand in the one at the top. When the last bit of sand has drained through, stop the clock and see what the time is.

If it only lasted about a minute, you need to carefully untape the bottle and add another 2fl oz of sand. If it's nearly reaching two minutes, just add a small amount. And if it's going far too slowly, take the tape off and use a pencil to make the hole in the disk slightly bigger. Now tape the bottles again and retime. Keep adjusting like this until you have the perfect two-minute timer—and the tool you need to keep your parents on their toes.

If you haven't caused enough mayhem yet:

Try making a series of timers with different **diameter** holes to record different time periods.

 ## The Sciencey Bit

Sand clocks can tell time consistently because sand has small, rounded **particles** of similar sizes which means the rate at which they move past each other stays the same. The rate of flow in a sand clock, however, can be changed by the **diameter** of the hole between the bottles.

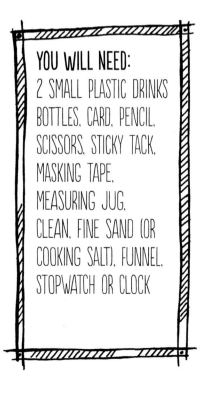

YOU WILL NEED:
2 SMALL PLASTIC DRINKS BOTTLES, CARD, PENCIL, SCISSORS, STICKY TACK, MASKING TAPE, MEASURING JUG, CLEAN, FINE SAND (OR COOKING SALT), FUNNEL, STOPWATCH OR CLOCK

Construct a liquid rainbow

Rainbows are lovely, but a little unpredictable, so if you want to have one on hand at all times, why not create a rainbow in a jar?

Begin by checking the volume of your jar. Do this by filling it with water and then pouring it into a measuring jug. Now work out how to fill it with five different layers by dividing this quantity by five—for example, if the jar holds 10fl oz, you need 5 x 2fl oz measures to fill it.

Next, pour out the correct measure of syrup into the jug and add a couple of drops of red and blue food coloring before mixing well with a spoon. You are trying to make purple, so if it's too blue, add more red, and if it's looking too red, add a little more blue.

Pour this into the bottom of the jar, and wash out your jug. Next, measure the same amount of blue dish soap or liquid soap and angle your jar so you can slowly tip it down the inside of the jar until it has formed a layer on top of the syrup.

Wash out the jug again and fill with a measure of water before mixing in a little green food coloring until it's a nice bright shade. Again, angle the jar so you can gently tip it down the side to form the next layer. Do the same with an equal amount of oil and then finally mix some red food coloring

into your rubbing alcohol and carefully add this before twisting the jar lid back in place.

Not only have you now managed to stack **liquids**, which is remarkable in itself, but you have also made a very impressive—and permanent— rainbow, perfect to have on hand for all those rainbow-based emergencies in your life.

If you haven't caused enough mayhem yet:

You could add another couple of layers. Try honey below the syrup layer—then attempt to color it purple while making the syrup indigo. And how about **experimenting** to see where milk fits into this **density** stack?

The Sciencey Bit

Everything is made up of tiny **particles** called **molecules**. Some of these **liquids** have lots of molecules in them and some have only a few even though the volumes of the liquids are the same. Those with lots (like the syrup) are more **dense**, and those with fewer, like the rubbing alcohol, are less dense, so the liquids form separate layers.

Tip: You could also add objects, such as ping pong balls or bottle tops, to see if you can find ones that will float in each layer.

Make your arms float away

First of all, you needn't be afraid: your arms may float away, but they will come back (which is lucky because they're very useful). Second, you will look a bit weird while doing this, so either explain to onlookers that you're in the middle of a classic science demonstration … or make sure no one is watching.

Begin by standing up, with your legs about shoulder-width apart and your arms hanging by your sides. Now grab a fistful of your trousers or skirt or whatever you are wearing with both hands and pull these outwards, really hard. And keep pulling, and pulling, and pulling. You should try to do this for as long as you can manage, but for a minute at least.

Now let go and allow your arms to float back down to your side and then after a second or two … whoa! Yes, that did just happen. Your arms floated up all by themselves. Weird or what?

Oh go on, have another go. You know you want to!

If you haven't caused enough mayhem yet:

Get a friend to hold their hands in a prayer-style pose, then place your hands over the top and push them together at the same time as your friend tries to pull them apart. Do this for about a minute then let go … and your friend's hands will magically pull apart.

The Sciencey Bit

Movement in your body relies on **muscles** which work by either contracting or relaxing. In this experiment your **triceps** muscles in your arms are contracting to try and move your arms up, but because you are stopping this happening the brain sends a message to push harder. The moment you let go, that message is still on its way from the brain and your arms move, even though you are no longer telling your muscles to tense.

YOU WILL NEED:
ARMS

Tip: Why not get all your friends to do this too—that way, there'll be no one left to laugh at you.

Create yogurt bites

Yes, that's right—yogurt "bites". I know that sounds wrong—like "milk chews" or "lemonade crunch"—but sloppy yogurt changes from a thick **liquid** to a **solid** when it is frozen (see, there are some science terms right there). It can also change from something a little dull to a fruit-tastic taste sensation (and no, in case you were wondering, fruit-tastic is *not* a science term).

First, lay out your fruit bite containers. For a large version, use silicone cupcake holders or paper cases laid in a muffin tin, or for mini bites, get out a clean, empty ice cube tray.

If you use cupcake cases, spoon in some yogurt until they are two-thirds full and then place on top blueberries, cut up strawberries or other fruit pieces, pushing them down a little with your finger.

If you are filling ice cube trays, mix together your yogurt in a bowl with finely chopped fruit, or simply squish soft fruit such as raspberries with the back of your fork before mixing it all in together. Now spoon the mixture into the ice cube compartments, making sure they are levelled off at the top.

Put your creations in the freezer for a couple of hours or until you can see they have changed into a solid. Now simply press them out onto a plate or bowl and dig in. Oh, and if you can't manage to eat them all, you can store them in a plastic bag in the freezer for a couple of weeks.

If you haven't caused enough mayhem yet:

Try creating your own recipes. You could add crushed nuts, dried fruit or edible seeds like sunflower, flax or sesame.

The Sciencey Bit

Temperature can cause things to change state. In this case, as it becomes much colder, the state of the yogurt changes from a thick **liquid** to a **solid**. If you allow it to warm up again it will change back from a solid to a thick liquid. This is a very good excuse to make frozen treats.

Tip: If you have lolly sticks you can try making frozen yogurt ice lollies as well.

Do the magic straw trick

This will amaze your friends! And if it doesn't, it might be worth finding a more gullible set of people. In fact, why don't you try it on some adults—that should work.

First, place the bottle on a table and balance your straw on its lid. Now invite your audience to try and move the straw without blowing it, or touching it—or moving the bottle, or the table (in case they're feeling really strong). They will, of course, fail.

Now you will show them that you can move the straw … using only your mind! First, let them examine the straw and pass it around. Then explain that you'll just make sure all the grease from their fingers won't affect things, and give the straw a good rub down with your top, or a dish towel, moving it back and forth *a lot*. And while your audience might think you're cleaning it, you are in fact charging it up with a load of static electricity.

Next, place the straw back on the bottle. It's best to do this by holding the middle of it in place with a finger then removing this in an upwards movement so the straw doesn't move.

Finally, astound everyone by putting your fingers on either side of your head in an 'I-am-about-to-use-the-awesome-power-of-my-mind' pose before letting your hands hover near the side of the straw. The electrical charge will be **attracted** to your hand when it's close enough and if you then move your hand away a little, the straw will begin to spin.

And while your friends may look at you in a suspicious "I-bet-this-uses-static-electricity" kind of way, any adults will be completely fooled. Guaranteed.

If you haven't caused enough mayhem yet:

You can also try using **static electricity** to make an aluminum drink can move. Simply charge up an inflated balloon by rubbing it on your top or hair and then hold this behind an empty can laid on its side and watch as it rolls away. You can even line up two cans, grab a couple of balloons, and have a race with a friend.

YOU WILL NEED: A BOTTLE WITH A FLAT-TOPPED LID, A STRAW, YOUR CLOTHES OR A DISH TOWEL

 ## The Sciencey Bit

You have just seen electricity in action! Rubbing with the cloth caused tiny negatively charged **particles** called **electrons** to pass from the fabric to the straw. These extra electrons gave it an overall negative electrical charge. Meanwhile your hand and the cloth lost some negative electrons in the process so had an overall positive electrical charge. As positive **charges attract** negative charges, the straw was attracted to your hand. When electrons flow freely they create an **electrical current**, but because these charges stayed where they were in the straw, cloth and hand, this is called **static electricity**.

Eat a rainbow

Here's the thing—although **light** looks white, it's actually full of colors. You don't see them very often but when the water in raindrops **refracts** sunlight it splits into seven different colors—red, orange, yellow, green, blue, indigo and violet—making a rainbow.

I can tell you this, and you might remember it. I could show you this, and you will probably remember it. But if I let you eat a rainbow, you will definitely remember it.

Lay out your cups in a line and add two tablespoons of milk to the first four of them, three tablespoons to two and leave the last one empty.

Now take your three primary food colors—red, blue and yellow—and begin creating your rainbow ingredients. Add six drops of red to the first cup, six drops of yellow to the second and finally nine drops of blue to one of the fuller cups, giving each one a stir (rinse your spoon in between so you don't mix colors by accident).

In the third cup, add three drops of red and three of yellow to make orange, to the fourth mix three drops of blue and three of yellow to make green, and to one of the fuller cups mix five drops of blue and five drops of red to make purple. Finally, in the empty cup add a tablespoon of blue milk and a tablespoon of the purple milk to make indigo.

Now you can paint your rainbow. Use a pastry brush or clean craft brush to add your milk paint in stripes across a slice of white bread from red, through to violet.

When it's done, ask an adult to put it in the toaster and when it pops up, spread on some butter before consuming a delicious rainbow (and all the knowledge this includes). What's that? You're not sure you've quite got it? Best make another slice then. Learning by eating—this could catch on!

If you haven't caused enough mayhem yet:

You can paint all sorts of pictures on your bread—not scientific but fun, artistic and tasty, so who cares?

The Sciencey Bit

Sunlight may look white—like this bread—but it's actually a **polychromatic light**. This means it is made up of many colors—red, orange, yellow, green, blue, indigo and violet. The bread only looks white because it is **reflecting** back all these colors into your eyes at once, which combine to appear white. However, when you paint red food coloring onto the bread this only reflects the red light into your eye and **absorbs** the other six colors. This is why it looks red. Blue coloring only reflects blue light and green only reflects green light and so on.

Tip: Make sure you don't use too much "paint" or your bread will get soggy.

YOU WILL NEED: 7 CUPS, MILK, FOOD COLORING (RED, BLUE AND YELLOW), PASTRY BRUSH OR CLEAN CRAFT BRUSH, WHITE BREAD, TOASTER (AND AN ADULT), BUTTER, KNIFE, PLATE

YOU WILL NEED: BROWN PAPER, PENCIL, SCISSORS, CARD, GLUE, HOLE PUNCH, COLORED PAPER, ENVELOPE

Create a weather tree

Come on! Hurry up! If you start this picture today, you'll be finished in … a year's time! Yes, you did hear me right. This project will take a whole year to complete, but it'll be worth it, I promise (and if it's not, I've got 365 days to think of an excuse).

Begin by creating your tree shape. Do this by cutting out a long, thin, triangular shape from some brown paper and sticking it onto a sheet of card to make the trunk. Now cut 12 smaller strips of paper to form the main branches—one for each month—and then some even smaller strips to add to these as twigs.

Take a hole punch to make your circular "leaves". These will be in different colors to represent the various weather conditions. So that you don't forget which is which, you can make a 'key' on another piece of paper to show what each color means. For example:

Cloud = grey, Sun = yellow, Wind = green, Rain = blue, Snow = white

You can also glue an envelope to this second piece of paper where you can keep lots of leaves ready to use.

Now at the end of every day you just select the leaf that represents the main weather conditions for the last 24 hours. Then add a tiny dot of glue to the back, or rub it on the top of a glue stick, and fix it next to the branch and twigs of the month you're in. You can even write the date on the leaf—if you're good at tiny penmanship.

So have you done today's weather? Good— only 364 to go …

If you haven't caused enough mayhem yet:

Try setting up other weather trees to look at specific trends. If you have a thermometer you could use your different-colored leaves to record the highest daily temperature. Or why not set up a rain gauge and show how much rain fell by using different colored leaves to represent different amounts?

The Sciencey Bit

The weather tree and key will help identify changes in **weather patterns** over different months and **seasons**. Seasonal changes happen because the earth is tilted as it makes its annual journey around the sun. This means that at any one time parts of the earth will be leaning towards the sun and that area will receive more light— in what we call summer—while parts that are leaning away from it at that time receive less light and are in winter.

Draw up your own height chart

You know when your mum buys you an enormous coat that drags on the floor and says "Don't worry, dear, you'll grow into it"? I'm afraid she's right. In fact, you're growing all the time, but it's hard to spot (especially when you're bundled up in that enormous coat).

But if you want to prove to yourself that you really *are* getting bigger, why not create your own personalized height chart?

First, stick together your pieces of paper lengthways, overlapping each by about 1–1½ inches and taping the join on the back. It's important to make sure the edges line up, so butt them up against a ruler while you do this. When they are all done, you can use a glue stick to paste down the flaps on the front.

Now take your ruler and mark off every ½ inch, but keep checking it against a long tape measure to make sure it's accurate. If you mark up both sides of the paper, you can use the ruler to join these up and create a line.

Next, write in the measurements along the left-hand side and then feel free to decorate—maybe by adding colors or even including some other measurements of interest. If you're a history buff why not record the heights of famous figures from the past? Mad on soccer? Then note how tall your favorite players are—or even the height of the world cup (14½ inches, in case you're interested). Or if you're into dinosaurs, you could include the size of the largest T. rex tooth (11¾ inches), the length of the tiny Compsognathus dinosaur (30 inches) or even the **diameter** of a stegosaurus' brain (1¼ inches—yes, not the brightest).

Finally, attach your chart to the wall with sticky tack and, making sure the paper starts at ground level, stand barefoot with your back to it while someone records your height (best done by holding a book level on top of your head). Write your name and the date alongside the measurement and do the same every three months to see if you can record a change. You never know, by the time you're 18 you might even be tall enough to fit into that coat.

If you haven't caused enough mayhem yet:

You could also use bathroom scales to measure your weight and record this too. And why not record the heights of the rest of your family? It's best to avoid asking for your parents' weight though—they can get touchy about such things.

 **The Sciencey Bit**

Human growth is usually measured by height as this records changes in a single system—the **skeleton**. Final height is influenced by things like **diet** but also determined by **genetics**—these are the **characteristics** you inherit from your parents. However, you will probably be taller than the previous generation as diet and health have improved over time, making you more likely to fulfil your maximum height set by your **genes**.

YOU WILL NEED:
PIECES OF PAPER,
STICKY TAPE, RULER,
GLUE STICK, PENCIL,
LONG TAPE MEASURE,
COLORED PENS OR
PENCILS, STICKY TACK

Make a bee and butterfly watering hole

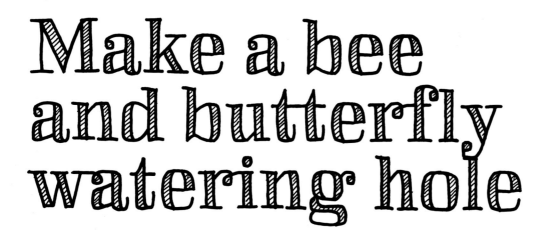

Yes, that's right, insects get thirsty too! You shouldn't be surprised—bees can visit thousands of flowers in a single day, so no wonder the poor things need a drink. It's not always easy to come by though. Puddles can dry up, and deeper water means they risk drowning, so it's a good idea to set up some watering holes around the garden.

First, you'll need a shallow bowl or saucer—even a lid. Next, add some pebbles so the tops of these sit just a little higher than the edge of your container. Now place this on the ground—somewhere near plants which you know bees, butterflies and other useful insects love. If you're not sure which these are, then go outside on a warm day and sit and watch the bees and butterflies to see which flowers they visit most (you could even take a nice cool drink of something—just make sure the bees don't get jealous).

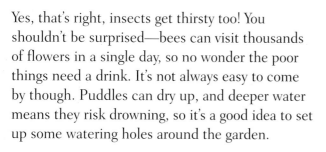

Now pour in water until it reaches the top of your container. You should notice that the tops of the pebbles act as islands—the perfect landing spots for insects. And why not add a few of their favorite flowers? This will help attract the insects, plus it looks good (which isn't exactly scientific but never mind).

You should check on your watering holes daily and top them up when they need it. And while you're at it, make sure you top up your own drink—okay, so you might not be flying at high speed through the air, but this nature watching can be pretty thirsty work, too.

If you haven't caused enough mayhem yet:

You could turn this into a feeding station by dissolving a little sugar in your water. Does this alter which insects use it and how often they turn up? Do temperature and weather conditions make a difference to the number of visitors?

The Sciencey Bit

Setting up watering holes is a great way to observe many garden insects. Insects belong to different families depending on features such as the number or size of wings they have, and you can use an identification guide to see how many different types of insects you can spot. You can also identify different **species** within a family, so you could record the number and name of different bee or butterfly species that visit your watering hole.

YOU WILL NEED: SHALLOW BOWL, SAUCER OR LID, PEBBLES, WATER

Tip: Why not have a go at using different metal objects such as metal coat hangers, whisks, or tins?

Conjure up magic bells

YOU WILL NEED: METAL SPOON OR FORK, STRING

People love the sound of bells, but seem less keen to have 400lbs of cast iron swinging about in their house. Thankfully, you can get much the same sound from a simple fork or spoon—as long as you're prepared to look a bit odd.

Grab a metal spoon or fork and tie it to the middle of your piece of string. Just make sure it is all metal and doesn't have a handle of a different material.

Next, wrap the other ends of the string a few times around the forefingers of both hands. Now gently swing the spoon or fork against something like a table or chair (not a person as they tend to object to having cutlery swung at them)—and listen. Not very impressive, is it?

Now do it again but this time hold your hands firmly against the side of your head so the fingers are next to (but not inside) your ears. Amazingly, you'll hear a sound that will make you think you're listening to an enormous ringing bell in a beautiful and impressive bell tower ... not standing in your kitchen looking a bit weird with a fork dangling from your hands.

If you haven't caused enough mayhem yet:

See what happens if you use non-metallic objects and also experiment by swapping the string for another material such as yarn, wire, or even rubber bands joined together.

The Sciencey Bit

You hear sounds because vibration of air against your eardrum stimulates your **nerves**, which then send electrical signals to your brain. When the string is next to your ears, these vibrations travel along the string and have less opportunity to spread out, meaning more of them enter your ear. This causes your eardrum to bend backward and forward with bigger movements making a louder sound (also called a high **amplitude**).

Levitate a ping-pong ball

Most adults think hairdryers should be used to dry hair. This shows a shocking lack of imagination. In fact, they are much better employed making ping-pong balls float in mid-air.

Before you begin, you need permission to use a hairdryer—oh, and you had also better check no adults want to dry their hair. If they do you could suggest they simply shake their head vigorously from side to side instead—after all, it works for dogs.

With the hairdryer plugged in and pointing upwards, place a ping-pong ball on the nozzle. Finally switch it on and watch as the ping-pong ball rises up and then floats in mid-air. If you're having difficulties, get someone else to place the ping-pong ball in the air stream so you can concentrate on holding the hairdryer still.

If you haven't caused enough mayhem yet:

Try to see how long you can keep the ping-pong ball in the air—and then challenge someone else to beat your record.

The Sciencey Bit

Usually the ball would drop to the ground because of the **force** of **gravity** pulling it downwards. But the air from the dryer pushes the ball up and the point where the two forces balance each other out is where the ball stays. Fast moving air also leaves behind it low **air pressure**, so the stream of air coming from the dryer forms a column of low pressure which the ball sits inside—and the higher air pressure outside makes it more difficult for the ball to move out of the stream.

Tip: To begin with, use the lowest setting on your hairdryer. You should also use the coolest setting as it is safer.

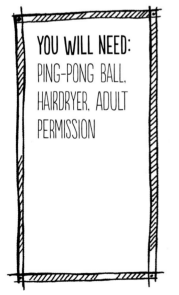

YOU WILL NEED:
PING-PONG BALL,
HAIRDRYER, ADULT
PERMISSION

Plant pollinator pots

If you want more bees, butterflies, or even moths visiting your garden, you could try putting up some welcome signs or have cut-price entry tickets for anyone with wings. But you'll find it a lot easier to simply grow pots full of their favorite plants.

You can make a recycled pot out of almost anything, provided it won't fall apart when left outside, and you are able to add holes in the bottom so extra water can drain away. If you can't snip or puncture holes yourself, ask an adult to add some using a hammer and nail or a drill.

Some things have got too many holes (like colanders or strainers) or one enormous one (like tires) and with these you might want to add a plastic lining with a few holes snipped in it (old potting soil bags work well).

Now you need to place your pots somewhere sunny and fill them with potting soil before sowing **seeds** of plants which will attract different **pollinating** insects.

Butterflies—try *Verbena*, scabious (*Scabiosa*), *Ageratum*, marigolds (*Tagetes* and *Calendula*), Sweet William (*Dianthus barbatus*), stocks (*Matthiola*), wallflowers (*Erysimum*), *Zinnia* and candytuft (*Iberis amara*).

Bees—sow borage (*Borago officinalis*), Californian poppy (*Eschscholzia californica*), *Cosmos*, cornflower (*Centaurea cyanus*), love-in-a-mist (*Nigella*), *Heliotropium*, poached egg plant (*Limnanthes*), sunflowers (*Helianthus*), *Phacelia*, *Cerinthe major*.

Moths –grow sweet rocket (*Hesperis*), night-scented stocks (*Matthiola bicornis*) and tobacco flower (*Nicotiana alata*)—these are all evening scented, and that is when most moths are active.

Keep the pots well watered in dry weather and within a few weeks you should have flowers in bloom and bees, butterflies, and moths queuing round the block to visit (actually, they don't queue—no manners at all, those insects!).

If you haven't caused enough mayhem yet:

Try painting some of your pots. Use an exterior primer and when it's dry, overcoat in acrylic or exterior paints—even chalkboard paint can work.

The Sciencey Bit

These flowers have **evolved** alongside the bees, butterflies and moths and rely on these insects to **pollinate** them by carrying **pollen** between the male and female parts of different flowers so they can produce fruit and set **seed**.

YOU WILL NEED: A RANGE OF OLD CONTAINERS, POTTING SOIL, SEEDS, WATER, WATERING CAN

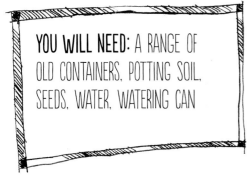

Tip: If the seedlings become too crowded, take some out to give the others room to grow.

Create a fall leaf window tree

Fall leaves are great but never seem to last long enough. One day there are enough for you to kick them all the way to school, or throw a handful over your little brother's head, the next they're a sludgy mess on the side of the street.

Thankfully, it's easy to capture a few of your favorites for months to come because the tree has done a great job at preserving them.

Yes, the **bacteria** and **fungi** which would break down these leaves need water to grow and reproduce, but the tree **absorbs** all the water it can from the leaf before it drops. It also absorbs the green **pigment** called **chlorophyll,** which is why the leaves no longer look fresh and green. Instead, you can see the waste products (brown) or the trapped sugars (red) which are left behind.

So, pick some of these lovely leaves from the ground, but try and select ones which don't show any damage, cuts or spots as these won't keep as well. And as pretty as the red and yellow leaves look, they don't hold their color as well as the browns, so choose wisely!

Next, lay your leaves out on the counter and make sure they are fully dry before peeling off the protective layer of your clear contact paper and laying the leaves onto it with spaces in between. You can then lay a second piece of clear contact paper on top and press them firmly together, getting rid of as many air bubbles as possible.

Alternatively, if you have a laminator you can seal them in laminated sheets or use an iron on a low heat to trap them between two sheets of wax paper. When you're done, cut around the leaf shapes leaving a slight margin where the plastic or wax paper is sealed.

Now cut out a tree trunk and branch shapes from some brown colored card and attach them to the window with small pieces of sticky tack or double-sided tape. Finally, attach your preserved, sealed-in leaves to complete your not-going-anywhere-in-a-hurry fall tree.

If you haven't caused enough mayhem yet:

Try to identify which type of tree your leaves came from and then make a note of which leaves last the longest on your window tree.

 ## The Sciencey Bit

Deciduous trees drop their leaves in fall. This is because in winter they would continue to lose water through their leaves but couldn't replace it via their **roots** as most water in the ground would be frozen. Before shedding leaves, a tree extracts water, **chlorophyll** and other useful materials as well as adding into them waste products it no longer needs—rather like the tree taking an annual visit to the bathroom. When spring comes around, the tree starts afresh by growing new leaves.

YOU WILL NEED: FALL LEAVES, CLEAR CONTACT PAPER OR LAMINATOR SHEETS AND LAMINATOR OR WAX PAPER AND IRON, SCISSORS, BROWN CARD OR PAPER, PENCIL, STICKY TACK OR DOUBLE-SIDED STICKY TAPE

Make your own sound effects

If you want to add a bit more drama to your stories there's nothing like throwing in a few sound effects. And although you might not have a Hollywood budget, you can create some pretty impressive noises without even dipping into your piggy bank (unless you want the sound effect of someone dipping into their piggy bank, of course).

You can experiment with lots of different objects and materials around the house but here are a few good ones to try:

• twisting or crinkling cellophane (crackling fire)

• squeezing a box or double bag of cornstarch(footsteps in snow)

• blowing through a straw into water (boiling water)

• snapping open an umbrella (sudden ignition of fire)

• sprinkling rice, birdseed or coarse sand on a metal baking sheet (rain)

• spinning a bicycle wheel and then letting a cloth rub against it (wind)

• squeezing folded sandpaper (breaking eggs)

• squeezing water bottles into a bucket (milking a cow)

• knocking together two halves of a coconut shell (horses' hooves)

• tapping large metal pot lids while holding them by the handle (bells)

• waving a pair of leather gloves up and down (bird's wings flapping)

• opening and closing an umbrella very quickly (bats flying)

Why not gather up lots of materials and practice making your sound effects "live" while telling the story (or getting someone else to)? You could even ask the audience to close their eyes to get the full effect.

Otherwise, you can record your effects to play when needed. If you go for the recorded option, try to do this in your own studio. Okay, when I say studio I mean find a small, quiet room, preferably with carpet and curtains plus lots of extra cushions and blankets. All these help **absorb** the sound waves from any extra noises so you'll be able to record just the sound nearest the recording device (which might well be the sound of you snoring as you get a bit too comfy on a bed of cushions and blankets).

If you haven't caused enough mayhem yet:

Challenge yourself to make as many sound effects as you can using simple everyday objects.

 ## The Sciencey Bit

We hear sounds because vibrations that are caused by the source of the noise enter our ears. Sound effects work because the **sound waves** they create are a similar speed, **amplitude** (loudness or quietness) and wavelength (**pitch**) to the noise they are trying to mimic so our ears hear a similar sound.

YOU WILL NEED:
HOUSEHOLD
OBJECTS,
RECORDING DEVICE
(OPTIONAL)

See your own DNA

The color of your eyes and hair, the sound of your voice and really important things like whether or not you can wiggle your ears or roll your tongue—all of this information is carried in every single **cell** of your body in a long **molecule** called **DNA**.

Amazingly, you can actually see your own DNA. It just requires a little bit of science, some mouth washing … and a steady hand.

You are going to need some rubbing alcohol—and if you ask an adult to put this in the freezer for a few hours first it will be a lot more effective.

Start by taking a swig of bottled still water and swill, swirl and squelch it around your mouth for a couple of minutes. You want to get as many cheek cells as possible into this so it can help to gently pull your teeth over the inside of your cheeks too.

When you have finished, spit the water into a clear plastic cup or glass and add ¼ teaspoon of salt, stirring it in until it's dissolved—this helps break down the cell membranes to release the DNA.

Leave this to sit for a minute or two and then add a drop of dish soap and stir again, very gently—you want to mix it in but without creating bubbles. The molecules in the dish soap hold onto the fats in the cell membrane, moving them out of the way of the DNA.

Finally, take your ice-cold rubbing alcohol, pour about 1fl oz into a measuring jug and then tilt your glass or cup and gently pour in the alcohol so it dribbles down the side of the glass and forms a layer above your **gene**-infested water.

Now watch closely. In a minute or two you will start to see white cloudy shapes appear and then some stringy substances. This … drumroll please … is your DNA! Yes, it's the information that makes you uniquely you. This should be revered, marvelled at … and then probably poured down the sink before somebody uses it to clone any more of you. One is quite enough, thank you.

If you haven't caused enough mayhem yet:

You can try extracting other people's DNA—or even DNA in fruits like strawberries or bananas. Squish these with a little water, salt and a drop or two of dish soap until it's a squidgy liquid, then pass this through coffee filter paper into a clear plastic cup before adding the ice-cold rubbing alcohol as you did before.

The Sciencey Bit

Surgical spirit is made of alcohol. Because **DNA** doesn't dissolve in alcohol it will start to precipitate (form lumps of **solid** in the **liquid**) out of the cheek **cell solution** when it comes into contact with the alcohol. Although a single strand of DNA would be far too tiny to see, in this experiment thousands of these strands are clumped together which means they become visible. Amazing!

YOU WILL NEED: RUBBING ALCOHOL, BOTTLED STILL WATER, CLEAR PLASTIC CUP OR GLASS, TEASPOON, SALT, DISH SOAP, MEASURING JUG

139

Build a snack delivery zip line

I have a question. What is better than a chocolate cookie?

A chocolate cookie delivered by zip line obviously! I can't believe you didn't get that. You really are slipping.

I guess it's time to improve your poor knowledge of snack delivery systems. To start, you'll need to decide where you want to run your zip line. You can set this up inside or out—it really depends where snacks are needed most.

Tie your string between two points. The important thing is that it runs from high to low, so **gravity** can get things moving. Next, you need to build your snack-carrying device. You could use a plastic or paper cup or perhaps a small plastic container. Whichever you decide, you'll have to find a way of hanging it from the string. You could try attaching wire or string around or underneath the container with sticky tape—just make sure it balances when it hangs otherwise the cup or container might tip out the cookies and that would never do.

You could then loop the string holding the carrier over the zip line or bend the wire into a hook shape to hang it on. If you do this you may find that it won't move far, if at all, because the string

or wire rub against the zipline, causing **friction**.

Try reducing friction by cutting a section of straw, splitting it lengthways, and then hooking it over the zip line, before taping your hanging container to it. You could also try making the string smoother by coating it in some oil.

And, of course, you will need to check if the number, shape or weight of the snacks you are transporting make a difference to the speed of delivery. If you find there are too many snacks for the system to work, you may even be forced to eat some—science is hard work like this sometimes.

If you haven't caused enough mayhem yet:

Challenge a friend to make their own snack delivery vehicle and use a timer to see whose arrives fastest along the zip line. Or try building larger carriers to transport toys. Experiment with different zip line materials, such as fishing line or yarn, to see which is most effective—can you predict which is best?

 ## The Sciencey Bit

Friction is a force that slows down touching objects that are moving over each other. Oil lubricates the surface of the wire and stops this and the container hook rubbing against each other as much. Smoother surfaces, like the straw, also create less friction so the movement of the container is faster.

YOU WILL NEED:
STRING, CUPS OR PLASTIC CONTAINERS, STICKY TAPE/ DUCT TAPE, STRAWS, SNACKS, WIRE (OPTIONAL), COOKING OIL (OPTIONAL).

Tip: If you tape an extra-long string to the base of your container you can pull it back up to the start of the zip line without having to move.

Make a floral suncatcher

If you're called an "attention seeking show-off" it's usually a bad thing—unless you are a flower, of course. Then the only way to respond is "Thank you—I try my best!" and wait for the **pollinating** insects to notice you.

Sadly, these bright, colorful flowers aren't usually around for long, but you can capture a few of them to keep. To do this, pick your flowers (make sure they are dry) and lay them carefully between two plain pieces of paper. Now slip these between a couple of sheets of newspaper, before sandwiching the lot between some heavy books. And wait … for a couple of weeks (sorry!).

Next, cut the center out of two paper plates (bend in the middle and snip to start you off) and then draw around this inner circle twice onto the back of your clear contact paper. Cut these out but make sure you add about ½ inch around the edge so it's bigger than the paper plate hole.

Peel off one circle and stick it to the back of a paper plate so the sticky side is showing through the center. Now you can get artistic and arrange your pressed flowers for the maximum impact. If you get stuck, pretend you're a flower—they're great at creating a showy look.

When you're happy with the design, peel the back off the other circle and place it, sticky side down, on top of the flowers, sealing them in. Smooth this out and then add glue around the paper plate edge before sticking the other plate on top to create the frame.

Now you can either use sticky tack to attach your suncatcher to the window or use a hole punch to create an opening through which you can thread yarn, string or ribbon to hang up your suncatcher. As the **light** shines through, it will illuminate the colors—and possibly confuse a few bees who might try visiting.

If you haven't caused enough mayhem yet:

You can try making a different suncatcher each month with flowers in bloom at that time. And why not use your pressed flowers to decorate placemats or coasters cut from card—just cover them in clear contact paper to protect them.

Tip: You can paint or decorate the top paper plate to make a more elaborate or colorful frame.

 The Sciencey Bit

They are very pretty to look at, but have you ever thought about what **petals** really do? Well, they're actually specially adapted leaves that are very colorful because plants use them to attract pollinating insects like bees and butterflies. These creatures come to drink the **nectar** and gather **pollen** but when they fly from one flower to another they also transfer some grains of pollen between them, which **pollinates** the flowers. Pollinated flowers can then go on to form fruits and **seeds**.

YOU WILL NEED: FLOWERS, PAPER AND NEWSPAPER, HEAVY BOOKS, SCISSORS, 2 PAPER PLATES, PENCIL, CLEAR CONTACT PAPER, GLUE, HOLE PUNCH, YARN, STRING OR RIBBON

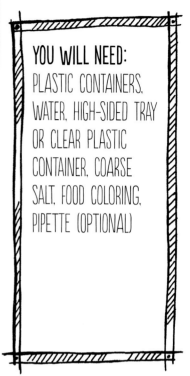

YOU WILL NEED:
PLASTIC CONTAINERS,
WATER, HIGH-SIDED TRAY
OR CLEAR PLASTIC
CONTAINER, COARSE
SALT, FOOD COLORING,
PIPETTE (OPTIONAL)

Tip: You can use a pipette to drop the food coloring more accurately.

Build a multi-colored ice sculpture

Ice sculptures look fantastic, but creating them usually involves people wielding hammers and chisels. Not surprisingly, parents will think this a teensy bit dangerous. Thankfully, you can create an impressive multi-colored **ice** sculpture with the help of something much less dangerous: salt.

First, you'll need to freeze your main ice block. Do this by placing a large plastic pot filled with water in a freezer overnight. The next day, a short while before you want your sculpture ready, place a high-sided tray or large clear container on the table and get your frozen block out of the freezer. If it won't turn out onto your tray straight away, leave it for a few minutes on the side and try again (the ice at the edges will have begun to melt slightly by then).

Next, sprinkle the top with some coarse salt and within a minute or two you should start to see the ice melting. You can now add color by carefully dripping some food coloring onto the top of your ice block. You should see it race along the melted crevices and form lightning-style bolts of color.

When you are happy with the effect, you can call in your fellow diners to eat around your impressive ice sculpture. In fact, it's probably best to continue the "ice" theme and feast on ice cream and slushies—perhaps with ice lollies for pudding.

If you haven't caused enough mayhem yet:

Try using fine table salt instead of coarse salt—what difference does this make and why do you think this is? Why not add some colored water to silicone-shaped cupcake molds or ice cube trays before popping them in the freezer? You can then use these colored shapes to decorate around the base of your ice sculpture.

The Sciencey Bit

Salt lowers the **melting point** of water. This means the **ice** starts to melt as salty water needs to be colder than this to freeze. It creates crevices in the ice block which the food coloring can then enter.

Make a shadow puppet theater

If you like plays with dark stories full of shadowy characters, this is the theater for you. In fact, there are nothing but shadowy characters here— even the good guys are downright shady.

To make your theater, find a good-sized box—but try to make sure the cardboard isn't too thick or it will be very hard work to cut. Now remove the front and back top flaps and then cut the front of the box out too. Flip it over so the two side flaps are splayed out as support legs and the back of the box is now facing you. Sketch out your theater design on this—you can make it as elaborate or plain as you wish—before cutting it out.

Cut a strip off one of the top flaps you removed and tape this along the bottom of the theater front to strengthen it. Tape greaseproof paper or tissue paper over the opening from behind to create a screen (see diagram).

You can make your characters and scenery by hand drawing them or by tracing the outline from pictures and then transferring them onto card and cutting out the shapes. Scenery can be attached to the sides and base on the inside of the theater and puppets should be taped to straws, pencils or skewers.

When you have everything ready, place your theater on the edge of a table and set an uncovered lamp or strong flashlight on another table about half a metre behind so it shines at the screen. Now you can start your show by bringing in your shadowy characters. The nearer the screen they are, the smaller they will be, and the farther back they go, the larger they appear. And just remember, dark tales and black mysteries always go down well, as long as your hero sees the light in the end.

If you haven't caused enough mayhem yet:

Try making jointed puppets by using a split pin to join two sections taped to separate straws, pencils or skewers.

The Sciencey Bit

Shadows are the absence of **light** and, as light travels in straight lines called rays, when you hold up the scenery or puppets you block the light rays in the same shape. If you hold the shape nearer the lamp, you are blocking more of the light rays so the shadow is bigger.

Tip: If you want multiple scenery changes, you can cut sections out of the side of the theater and slide different sets in and out.

Making the theater

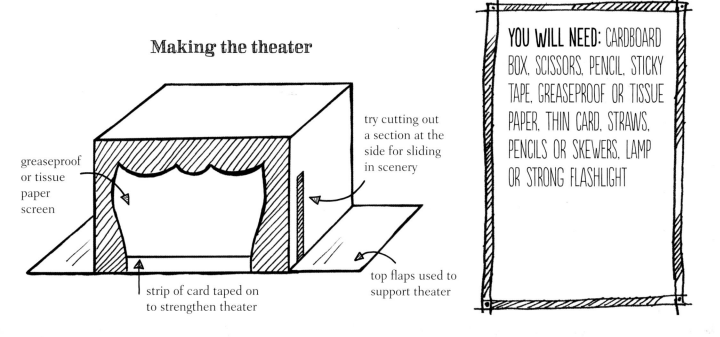

greaseproof or tissue paper screen

try cutting out a section at the side for sliding in scenery

strip of card taped on to strengthen theater

top flaps used to support theater

YOU WILL NEED: CARDBOARD BOX, SCISSORS, PENCIL, STICKY TAPE, GREASEPROOF OR TISSUE PAPER, THIN CARD, STRAWS, PENCILS OR SKEWERS, LAMP OR STRONG FLASHLIGHT

Tip: If your container has rough edges you can sand them down or add some padding to your design.

Launch water bombs

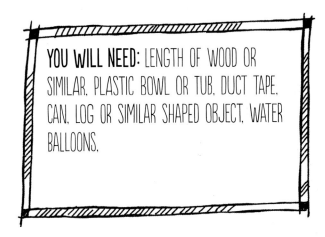
You know when you're having a water balloon fight and someone is too far away to be hit, or keeps avoiding your shots? And they're standing there, pulling faces and generally looking a bit smug? Well, this is where your water balloon launcher comes in. After all, if science can't help you get your friends and family soaking wet then what use is it?

First, you'll need to find a **lever**—this should be an object that is long and strong and will sit nice and flat, such as a length of wood. Next, you'll need something to place your balloon in. A round plastic bowl works well but you could also try different containers to see which is most effective. Just make sure there are no sharp corners for the balloon to catch on or the only person getting wet will be you. Attach this container to one end of your lever using duct tape and then find a **fulcrum**. This is an object, like a can or a log, on which you can rest your lever.

Finally, you need to test out your launcher. Place a full water balloon in the container and then step quickly on the other end of the lever. The nearer the center your fulcrum is, the farther the projectile (balloon) will fly, but the nearer the container it sits, the more **energy** you will transmit so the higher the balloon will go (useful to catch people unawares). Whichever way you look at it, no one's going to be looking smug for long.

If you haven't caused enough mayhem yet:

Try moving the fulcrum to different positions and record the effect this has in terms of height and distance. If you mark these on the side of the wood you will be able to target people more effectively.

The Sciencey Bit

When someone uses a **lever**, they exert a **force** (the effort) around a **pivot** (the **fulcrum**) to move an object (the load). By moving the pivot you can reduce the force you need to lift a load, or launch a water balloon.

Put someone in a cage

Yes, that's right, you can put your sister* into a cage using something called a **thaumatrope**. Unfortunately, this is an optical illusion as it's not a real cage. But if it cheers you up, you could also have her eaten by a monster, swim in eel-infested waters or become a chicken. See, I knew that would make you feel better.

To make your thaumatrope disk, draw around a cup on a piece of thick card and cut it out. If you only have thin card, cut out two or three disks and stick them together with glue.

On one side of the card disk you can draw a picture of your sister*—or if you have a spare photo, you can use this instead; just make sure it fits onto the disk shape. Now flip it over and draw the bars of a cage on the other side.

Use the hole punch to make holes on each side of the disk and then cut two pieces of string 20 inches long. Now thread a piece of string through each hole so you have a double string on either side (see diagram).

You can then wind the picture around but it's quicker if you hold the ends of the string and swing the picture over again and again as if you were using a skipping rope. When you can see lots of twists down the string, just start moving your hands apart. As the string untwists and flips around the pictures, they will blend into one and you will see your sister trapped behind bars—where she belongs!

*This trick is not limited to sisters. You can humiliate anyone—it's very democratic like that.

If you haven't caused enough mayhem yet:

Forget the bars; try adding all sorts of scenarios on the other side. And you could try other, non-sister-humiliating designs too. Then again …

Making the thaumatrope disk

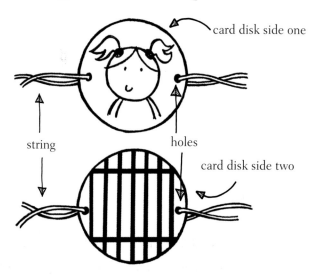

card disk side one

string

holes

card disk side two

👓 The Sciencey Bit

The **retina** is the part of your eye that changes the image you see into electrical **nerve** signals to be sent to the brain. But images stay on the retina for up to a tenth of a second—something called **retinal retention**—and as the **thaumatrope** spins it switches images faster than this, tricking your eye into seeing two separate images as one.

YOU WILL NEED: CUP TO DRAW AROUND, THICK CARD OR THIN CARD AND GLUE, SCISSORS, PENS/PENCILS, HOLE PUNCH, STRING

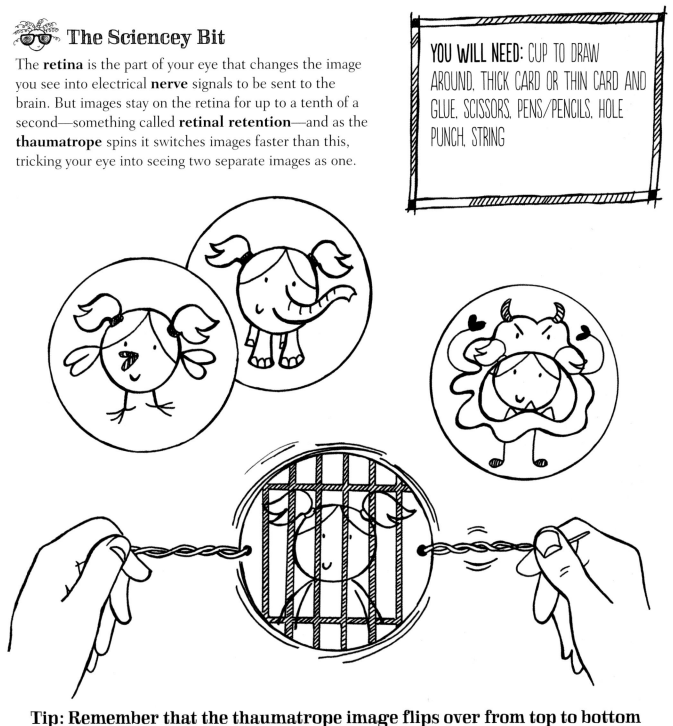

Tip: Remember that the thaumatrope image flips over from top to bottom so make sure the image on the other side is upside down.

Tip: If you are planning on washing your dyed materials afterwards, try to do this with a gentle detergent and separately from other clothes.

YOU WILL NEED:
NATURAL MATERIALS SUCH AS TEA, TURMERIC AND BLACKBERRIES, PANS, PLASTIC GLOVES, BOILING WATER (AND ADULT), WHITE MATERIAL, RUBBER BANDS, PROTECTIVE CLOTH OR NEWSPAPER, SIEVE

Try some natural tie dying

There are loads of natural dyes all around us. If you want to discover how many, just put on a white T-shirt before dinner and count how many times your parents scream "Don't get that on your top!"

To try staining things in a more *controlled* way you'll first need to extract some dye. Good natural materials to try are tea (brown), turmeric (yellow) and blackberries (purple). Take a pan for each dye, put on some plastic gloves and then add your material—5–6 tea bags, two teaspoons of turmeric or a couple of cups of blackberries. Next, ask an adult to pour in about 25fl oz of boiling water. Leave it for a half hour to stew so the color comes out and it can cool to a safer temperature.

Then you'll need to ask an adult for some material to use. That white T-shirt you've already got half a dozen stains on might work, or things like old sheets and pillowcases which you can cut up. Just make sure whatever it is has been washed and is wet. Next, you can twist, fold or scrunch the fabric up and then hold it in this position with rubber bands.

Next, make sure you protect all the surfaces. And then double-check everything is covered. And then check again. And then make sure you have your plastic gloves on. And then leave the house and do this outside. Sorry, I just don't trust you.

When you are well away from all precious objects, add your prepared material to the dye (you'll need to remove the tea bags or strain the blackberry dye through a sieve first). Let it sit there for at least half an hour and until you are happy with the color it's turning. Now take off the rubber bands and rinse off any extra dye in a bowl of cold water before hanging your material out to dry. Finally, check your own clothes are stain free … and breathe a sigh of relief.

If you haven't caused enough mayhem yet:

Experiment by dyeing with other natural materials. If the color doesn't stay, you can try pre-soaking the fabric in a salt **solution**—these are called **mordants** and can help the pigment **molecules** bond with the **fibers**.

The Sciencey Bit

Many plants contain natural ingredients that stain or dye things. Different **fibers** in different materials may form stronger or weaker bonds with the **pigment molecules** (the natural coloring matter in plants) so you can get darker or fainter colors with different fabrics.

Play a nature pairs game

Remembering the names of trees or differences between plants is not easy. Unless, of course, it means you have a chance of beating your brother in a game. Then it becomes a lot more manageable (and important).

Begin by measuring your card and then making a pencil mark halfway along the top and bottom before joining the two up with a line. Measure along the sides and again make a pencil mark at the halfway points—but also at the halfway points between the middle and ends before joining up with three lines (see diagram). Cut these out to make eight cards and then do the same with the other sheets so you have 32 cards in total.

Now you need to find 16 matching pairs to use on your cards. Look around your garden or outside on walks and then pick the leaves or flowers you spot—just make sure each pair is well matched (a similar size, color and shape) and will fit onto your card. Also, try to identify which type of plant or tree each leaf or flower pair comes from so you can write this on too.

Now you need to press your nature pairs to preserve them (see page 142) which means waiting a couple of weeks.

When they are ready, use a dot or two of white glue to attach each dried leaf or flower to a card and then write the name of the plant below. And to protect them during the rigours of playing, draw around a card on the back of some clear contact paper, cut this out, peel off the backing and use it to cover each one.

To play the game, shuffle the cards and then place them all face down. The first player can now turn over two cards—if they match, they can keep them and have another go. If they don't match the next player has a go. The winner is the person with the most pairs at the end of the game.

If you haven't caused enough mayhem yet:

Try making four of each type of card and you can use these in a game of nature snap as well.

The Sciencey Bit

You can recognize different plants by their flowers and leaves. When you are identifying them, look for clues like number, size and arrangement of **petals**, leaf shape and color, how the leaves are grouped together, hairiness of leaves and much more.

Tip: Continue to collect and press flowers and leaves throughout the year so you can make pairs games for different seasons.

Making the cards

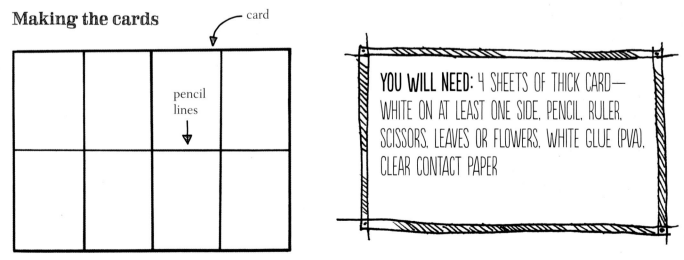

card

pencil lines

YOU WILL NEED: 4 SHEETS OF THICK CARD— WHITE ON AT LEAST ONE SIDE, PENCIL, RULER, SCISSORS, LEAVES OR FLOWERS, WHITE GLUE (PVA), CLEAR CONTACT PAPER

Make a lunar phase flipbook

This is the perfect project to do when days are quite long because you get to utter the following excuse "I would love to go to bed, but I'm afraid I need to stay up until it's dark to chart the phases of the moon. I believe it's waxing gibbous tonight."

They'll be so astounded that not only will they let you stay up late, you'll probably be given milk and a cookie too.

For this incredibly important astronomical work, you'll need a block of sticky notes, or a small pad of plain paper. Start at the front and work your way through the book tracing around a large coin on the bottom right-hand corner of each page. You'll need 30 pages to chart a complete lunar orbit.

You can wait to begin at a full moon or simply start straight away. You'll need to go out and see how much of the moon is in **shadow**, then shade this in on your first page. You can also write the date and what phase the moon is in (see diagram). Keep doing this every evening for a month until your book is full.

To see the phases of the moon come to life, you need to put down the milk and cookie (sorry) and then grasp the front right-hand corner of your book between your thumb and forefinger. Now let your thumb move slowly downwards to release the pages and watch the shadows move to reveal a moon moving through new, young, waxing crescent, first quarter, waxing gibbous, full moon, waning gibbous, last quarter, waning crescent and old.

If you haven't caused enough mayhem yet:

If this is too fast or confusing for parents, ask them to give you some more lovely round cookies and you can use these to demonstrate the lunar phases instead. You'll find it's a very tasty way to get from a full moon to a full stomach.

The Sciencey Bit

The moon is a sphere that travels once around the earth every 29.5 days. As it **orbits**, it is lit up from varying angles by the sun. When there is a "new moon", the moon is between the earth and the sun, so the side facing towards us receives no direct sunlight. As it moves around the earth, the side we can see gradually becomes more illuminated by direct sunlight. In a couple of weeks it is on the other side of the earth so the side we are looking at is fully facing the sun making a "full moon".

The phases of the moon

new

young

waxing crescent

first quarter

waxing gibbous

full

waning gibbous

last quarter

waning crescent

old

YOU WILL NEED: BLOCK OF STICKY NOTES OR SMALL PAD OF PLAIN PAPER, LARGE COIN, PENCIL, BLACK PEN

Tip: You can also transfer the information onto a lunar chart to hang on your wall.

Hang a twig mobile

It can be quite a skill to find your **center of gravity**—just ask a tightrope walker. Well, no, not when he's actually on the tightrope. In fact, instead of interviewing tightrope walkers, you might be safer finding the center of gravity with a twig mobile.

First, you'll need to find some twigs. The best place to look is at the base of trees where they may have dropped. Just make sure you never break a twig off a tree—those are the ones it needs. It's also good if you can find pairs of roughly matched twigs in a variety of sizes.

To add some color to your twigs, you can cover them in yarn. Simply tie on one end and then wrap. When you want to change color, cut off your first thread and tie it onto the next color and continue wrapping. For a neater look, try to trap the ends of your knots under the wrapped yarn.

You'll need to hang your mobile from somewhere while you work—a broom handle or pole balanced between the backs of two chairs works well. Start with a single large twig. Hold your thumb and forefinger in a circular shape and move the twig between them until you can feel it is balanced—this is the center of gravity and is the point at which you should tie a length of yarn. Attach the other end to the broom handle. If it's not perfectly balanced, nudge the yarn along the twig until it is.

Now take two smaller wrapped twigs and again find their center of gravity and attach some yarn at this point. Next, tie one onto one side of the first twig and one onto the other. If a twig pulls the mobile too far down, move it nearer to the center of the main twig where the downward **force** will be less. Likewise, if one side seems too light, move the twig farther outwards. You can keep adding pairs of twigs like this until you are happy with the look and feel so confident about finding the center of gravity that you're ready to take up tightrope walking.

If you haven't caused enough mayhem yet:

You can decorate the twigs further by hanging on ornaments. Light objects, such as feathers or hole-punched leaves, will be easier to balance, whereas heavier objects like pine cones will need to be carefully placed on either side to make sure the mobile remains level.

Tip: It helps to have a friend hold the twig steady whilst you add smaller twigs to either side.

The Sciencey Bit

The **center of gravity** is the point at which an object is perfectly in balance. Because the twigs are not a uniform shape and **mass**, this point will not necessarily be in the middle. You can also balance objects either side of the twig even if they have a different mass. This is because if you move an object nearer the hanging point, it exerts less **turning force,** and if it moves farther away, it exerts more turning force, allowing you to balance mismatched objects.

Tip: It can help to put more invisible "ink" on the paper by going over your message or drawings a couple of times.

YOU WILL NEED: BAKING SODA, WATER, BOWLS, COTTON SWAB OR PLASTIC "PEN" (E.G. PLASTIC CUTLERY, STYLUS), WHITE PAPER, RED CABBAGE, BOILING WATER (AND ADULT), SIEVE, PAINTBRUSH

Write with invisible ink

There are times when you might not want people to see what you're writing: when you're plotting ways to land your brother in trouble, designing your latest booby trap, or outlining plans to take over the world. And that's just off the top of my head.

Thankfully, this is where invisible ink comes in handy.

You can make your own ink by putting a couple of tablespoons of baking soda in a bowl and then stirring in a teaspoon or two of water.

Then you just need to create your message (or draw your dastardly booby trap design) by dipping a cotton swab in the "ink" and writing on some white paper. For a finer nib, try the end of a piece of plastic cutlery or any thin plastic implement. As the "ink" dries, your message will magically disappear, covering your tracks nicely.

Of course, you and your accomplices will have to uncover your brilliant plans to put them into action, which means you are going to need some "revelation paint". Make this by tearing up a couple of red cabbage leaves and placing them in a bowl. Then ask an adult to pour some boiling water over the leaves to release their magical purple juice. And to avoid throwing suspicion your way you may have to explain "I have a sudden craving for a delicious cabbage drink" with as much conviction as you can muster.

When the water has cooled, put the whole lot through a sieve so you're just left with the liquid in a bowl. Now use a brush to paint over your message and reveal your elaborate plans.

But just be careful—the cabbage juice can stain, so keep it off your clothes and anything precious, or the only thing you'll be revealing is how much trouble you're in.

If you haven't caused enough mayhem yet:

You can also try writing in lemon juice or white vinegar. What happens when you try the revelation paint on this?

The Sciencey Bit

Red cabbage juice acts as an **indicator**—chemicals in it (**anthocyanins**) react with the **alkali** or base (baking soda) and change from purple to greenish blue, revealing your elaborate plans. It will also react with **acids** like lemon juice and vinegar to turn a red color.

Create your own windmill

Windmills can be used to grind grains, pump water or even make electricity, but we're going to build one with a far greater purpose: to deliver candy.

First, you'll need the body of your windmill. You can recycle an old box, large paper cup or juice carton for this—just make sure it is thin enough for you to poke a kebab skewer through it from front to back and that this spins easily when it's in place.

Now to make your sails: cut a 8¼ x 8¼ inch square from an A4 piece of thin card and use a ruler to draw two diagonal lines from corner to opposite corner. Measure 1¼ inches back from the center point along each line and use scissors to cut from each corner to these points (see diagram). Next, place one corner triangle above a piece of sticky tack or playdough and push through a pin or tack to make a hole. Do the same for each alternating triangle as well as the middle point (see diagram).

Add a bead onto the pointed end at the front of the skewer, then thread on the center of your windmill before folding over and pushing on each corner that has a hole. You could also use glue to help stick these down. When all four are folded in, thread on another bead and then use a piece of playdough or sticky tack molded over the end of the skewer to hold the windmill's shape.

Finally, pierce a hole in a plastic or paper cup, thread through some string and tie it on before attaching the other end to the rear of the kebab skewer. Now, when the wind blows and the sails turn it will wind up the string which will, in turn, raise the cup carrier. Of course, that's not really putting your windmill to the test. To do this I advise seeing how many pieces of candy your windmill can lift in its paper bucket. Now there's a proper test.

If you haven't caused enough mayhem yet:

For a smoother turning windmill, try increasing the size of your holes so you can fit a straw between them for the skewer to pass through—this reduces **friction** and allows more **energy** to be used in turning the shaft.

The Sciencey Bit

The windmill is converting the wind's **kinetic energy** (the energy of movement) into mechanical power (which lifts the bucket). The windmill works because air is pushed into the sails which are specially shaped to make them spin around turning the shaft (the skewer), which in turn winds the string.

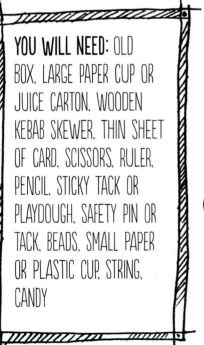

YOU WILL NEED: OLD BOX, LARGE PAPER CUP OR JUICE CARTON, WOODEN KEBAB SKEWER, THIN SHEET OF CARD, SCISSORS, RULER, PENCIL, STICKY TACK OR PLAYDOUGH, SAFETY PIN OR TACK, BEADS, SMALL PAPER OR PLASTIC CUP, STRING, CANDY

Making the windmill

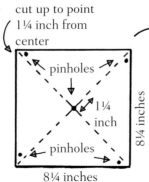

cut up to point 1¼ inch from center

pinholes

1¼ inch

pinholes

8¼ inches

8¼ inches

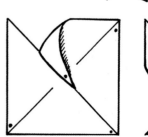

The Sciencey Bit

Solitary bees are important **pollinators**—this is because they fly between flowers gathering **pollen** and **nectar**. Pollen grains from the female part of the flower (the **stigma**) can get stuck on their furry bodies and one or two will rub off on the male part of the flower (the **anther**) when they visit the next flower—and this leads to **pollination** (see page 195).

YOU WILL NEED: TIN CAN, HOLLOW STEMS, SCISSORS, TWINE

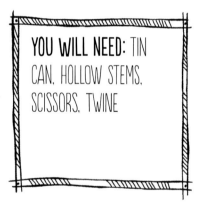

Tip: You can decorate the outside of your can with some acrylic paint.

Hang up a bee house

If you like strawberries, apples, cucumbers, blueberries and in fact a whole range of delicious fruits, vegetables, and nuts, then you owe a big thank you to bees. Without their **pollination** work, you wouldn't be enjoying any of these things. So why not build them a home (I did say a BIG thank you, after all)?

Don't worry, it's not as hard as it sounds. Although honeybees live in large hives with thousands of sisters and brothers, and bumblebees dig nests to house a few hundred, we're going to build a home for those that like to live alone—solitary bees.

First, find a clean, empty tin can and then look for some hollow stems of plants—the insides of which should be up to $1/3$ inch across. Try to get a good variety, as different sizes will attract a range of bees. Each one needs cutting to the length of the can with a pair of scissors. Bamboo canes also work well, but you may need to ask an adult for help cutting these as they are quite tough and need special tools.

Make sure you pack your stems in tightly so there is no chance of a troublemaking bird or squirrel pulling them out, and then tie a piece of string around either end of the can. Tie another piece of string to both these end pieces and you can use this to hang your insect shelter in the garden somewhere warm, sunny, and sheltered from wind.

Through the spring and summer you should see solitary bees taking up residence. You can look for telltale signs. No, there won't be removal vans parked outside, but if a bee has moved in and laid eggs, they will seal off the end of their individual stem home. What's more, the material they use will give you clues as to which bees are there. If it's plugged with leaves, it will be a leafcutter bee, with mud, it'll be a mason bee, with fine hair it will be a wool carder bee and with chocolate, it will be a cocoa bee.*

*Okay, I'll admit I made that last one up.

If you haven't caused enough mayhem yet:

Try making several bee houses using a variety of types and sizes of stems in each and hanging them facing in different directions and at varying heights. Which homes get the most occupants and why do you think this is? Try planting some bee-friendly flowers (see page 132) near the bee home and see what effect this has.

Create a solar still

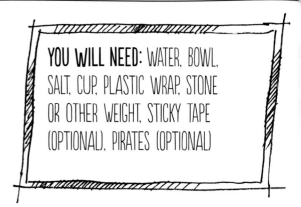
You are going to be grateful you know how to do this when you've been cast adrift by pirates for being too annoying and are floating on a vast ocean with only salt water to drink. What do you mean, that's far-fetched? Don't you know how annoying you can be?

So, back to this boat …

As you know, salt water isn't good to drink, which means you'll need to take the salt out if you are going to survive in your drifting boat. Thankfully, you can set up a simple solar still to extract it.

First, place some of the sea water in a bowl (well, yes, I do think the pirates would have given you a bowl) and then place an empty cup at the center (yes, I think the pirates might well have given you one of those too). Now cover the top of your still with plastic wrap (see diagram), and then place a stone or another sort of weight in the middle above the cup (let's just assume the pirates gave you a well-stocked boat, okay?). Now you just need to sit and wait … a long time.

Of course, if you are doing this at home you will need to make your own ocean water by stirring a couple of tablespoons of salt into the bowl of water. You can also go off and leave your still somewhere sunny for several hours while you do something else. However, in your boat you'll just have to sit there patiently wondering if you should have been a bit less annoying in the first place.

Once you have some water collected in your cup you can do a taste test—and it should be lovely fresh, salt-free water. If not, then that would be really … well … annoying.

If you haven't caused enough mayhem yet:

Try using different liquids, maybe orange juice or squash—what happens?

Making the solar still

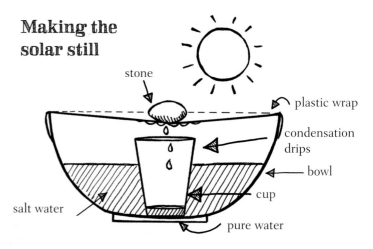

stone

plastic wrap

condensation drips

bowl

cup

salt water

pure water

 ## The Sciencey Bit

Gradually, as the water heats up in the sun, it **evaporates** to become a **gas** before rising and hitting the plastic wrap, where it cools and turns back into liquid water droplets—a process called **condensation**. Eventually, **gravity** makes the water droplets roll down the plastic wrap towards the rock where the water droplets slide off into the cup. The salt doesn't evaporate, so it gets left behind in the bowl.

Tip: If the plastic wrap doesn't stay in place easily, add some sticky tape on the sides to secure it.

Build a bottle tower garden

If you're a bit short on planting space in your garden, take a tip from skyscrapers and try building upwards. This way you'll have a space age tower block that's the perfect place to grow fruit, salad, and vegetable plants.

Take your first bottle and, with the lid still on, use the end of a safety pin to make small holes near the top—these will be for drainage. Now push and turn a ballpoint pen into the same holes to widen them.

Next, remove the cap and squash the bottle down so you can make a cut near the base and another halfway up. Pop it back into shape and then use the bottom slit to cut all the way around the base. Make a second slit a third of the way down the bottle about 2¾–3½ inches wide and add two 2-inch slits upwards before using masking tape to hold this closed. Pop the top back on, turn it upside down and fill it to within about 4 inches of the top with potting soil (see diagram).

Use wire to attach this first bottle to a fence or some trellis in a nice sunny spot and then add three or four more bottles on top (also attached with wire). These need their caps removed and no extra drainage holes, but they do need the base cut off and the slits added in the same way as you did for the first bottle.

To make a funnel for the top, just remove the base and lid of another bottle and turn it upside down and then add a final bottle with the base removed and a single hole made in its lid (see page 110 for instructions). Now put a couple of handfuls of coarse sand into this and you have your filling and filtering bottle to crown the top of your tower.

Make sure all the masking tape sections are facing the front, then remove the tape, bend down the flap and use your finger to make a planting hole. Add a small plant to each and then press back the soil around its base to hold it in place.

Finally, all you need to do is fill the top bottle with water and then let **gravity** and nature do the rest.

The Sciencey Bit

Plants need water to grow. They **absorb** this through their **roots** from damp soil. Pouring water too quickly into these towers would mean it gathered in the lowest bottle, getting those roots too wet, and leaving the higher roots dry. It could also wash out **nutrients** from the potting soil. By dripping water through slowly, the soil and roots at the top have more time to absorb the all-important water before **gravity** takes it downwards. The sand acts as a filter, sifting out large **particles** which might otherwise block the drip hole in the lid.

If you haven't caused enough mayhem yet:

Try building multiple towers—each could house a different sort of salad or vegetable. You could even try sowing **seeds** rather than adding small plants.

YOU WILL NEED: SEVERAL EMPTY 4-PINT PLASTIC BOTTLES WITH LIDS, SAFETY PIN, BALLPOINT PEN, SCISSORS, MASKING TAPE, POTTING SOIL, WIRE, FENCE OR TRELLIS, COARSE SAND, JUG OF WATER

Making the bottle garden

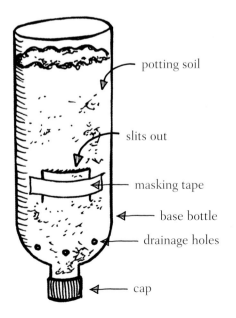

potting soil

slits out

masking tape

base bottle

drainage holes

cap

Concoct fizzing lemonade

Never mind space exploration, medical breakthroughs and industrial inventions, without science we wouldn't have the fizz in fizzy drinks. Yes, science is *that* important.

If you want to show off your own scientific prowess you can have a go at adding the fizz yourself. Begin by squeezing your lemon with the juicer and pouring it into the glass (if you don't have a juicer you can try squeezing by hand and passing all the pulp and juice through a sieve into the glass).

Next, add a teaspoon of baking soda to the lemon juice and give it a stir. At this point you can either think "Ah, I see I've caused a reaction by adding a base to an **acid** that has produced **carbon dioxide**" or you can just think "Oooohhh … look! Bubbles!"

Now add some water to the mix until you have double or triple the volume and then stir in a teaspoon of superfine sugar (to balance the sour taste of the lemon juice) and have a drink.

You should be able to see—and experience— the bubbles of carbon dioxide still in the mixture, and it's these that add the fizz to things like fizzy lemonade or cola. In fact this is why they're called *carbonated* drinks (I know—it's all starting to make sense now, isn't it?).

In fact, having made so many incredible discoveries you probably deserve a sit down, with a nice cool lemonade and some gelatinized starch products with added glucose (AKA cookies).

If you haven't caused enough mayhem yet:

Try refining the recipe to your taste. You can add more lemon juice to make it zestier, more water to dilute, additional baking soda for more fizz, or extra sugar to sweeten it. Try using different fruits as well—what difference does this make? Which work best and why do you think this is? Which cookies go best with lemonade (not entirely scientific, but never mind)?

The Sciencey Bit

When the lemon juice (**acid**) and the baking soda (**base**) mix, they create a **chemical reaction** known as an **acid-base reaction**.
This produces a **gas**, **carbon dioxide** (CO_2), which creates bubbles in the liquid in a process called **carbonation**.

Make a pinhole camera

I'll admit this camera isn't going to allow you to take any pictures, but you will get to turn the whole world upside down and that's pretty impressive, isn't it?

Begin by cutting down one of your cardboard tubes so it's about a third of its original length. Now measure both tubes and cut sheets of black paper to the same length, roll each up and insert them into the matching tube. They should then unfurl to give you two black-lined cylinders.

Next, cut a circle of tracing paper at least twice the **diameter** of the tubes, place it over the end of the longest tube and use a rubber band to hold it in place—this will be your **translucent** screen. Cut a circle the same size, but this time out of aluminum foil, and attach it to the end of the shorter tube, again with a rubber band.

Now join the tubes together with tape (see diagram) so the tracing paper screen is sandwiched between the two. You need to keep out as much **light** as possible, so roll some black paper or card over the join and hold it in place with more rubber bands.

And you're done. Oh, now hang on—you're just looking into a pitch black tube, aren't you? I think I might have forgotten something vital in this pinhole camera—the pinhole. So take a pin, tack or safety pin and make a pinhole in the middle of your foil end and then go outside and take a look through the tube. Magically, you should be able to see the view appearing on your tiny screen … but upside down.

If you haven't caused enough mayhem yet:

Think about how you could improve your pinhole camera, then try some of your ideas, for example, moving the screen nearer or farther away from the pinhole, using other materials for the screen or adjusting the size of the hole, or maybe just writing "World-upside-down-viewer TM" on the side.

The Sciencey Bit

Light **reflected** by objects passes through the pinhole and onto the screen. Because **light rays** travel in straight lines, those from the top of your view hit the bottom part of the screen as they pass through the tiny hole and those from the bottom of the view hit the top of the screen. They cross over as they pass through the pinhole and make the image upside down. This is also how your eye works—light passes through the **pupil** (your pinhole) and onto the **retina** (your screen) upside down. Thankfully, for your eye, your brain cleverly flips the image the right way up again.

172

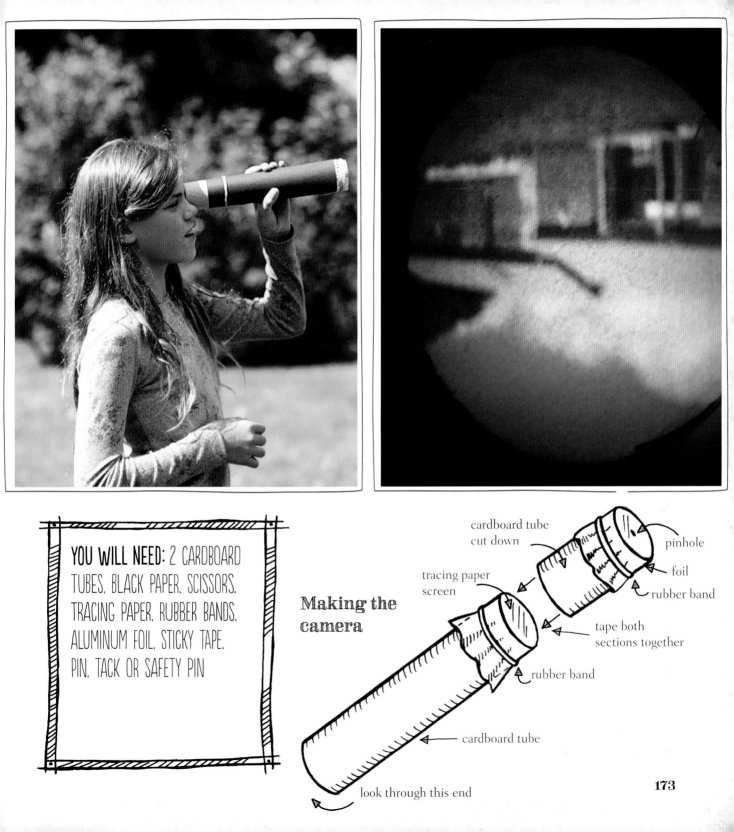

YOU WILL NEED: 2 CARDBOARD TUBES, BLACK PAPER, SCISSORS, TRACING PAPER, RUBBER BANDS, ALUMINUM FOIL, STICKY TAPE, PIN, TACK OR SAFETY PIN

Making the camera

cardboard tube cut down

pinhole

foil

rubber band

tracing paper screen

tape both sections together

rubber band

cardboard tube

look through this end

173

Watch an egg take a high dive

This is a spectacular trick if you get it right. If you get it wrong it is a spectacular mess. So no pressure there then.

First, get a glass and fill it three-quarters full of water. Now on top of this place something strong and flat like a piece of thick card, cake tin base or a coaster. Above this, stand a cardboard tube—a toilet paper inner tube works perfectly—but just make sure it's sitting directly above the center of the glass. Finally, lay an egg across the top of the tube.

Okay, now take off the raw egg and replace it with a lemon or a small ball or anything else that will function as a stunt-egg. I think we're both going to feel a lot better if you practice this first, right?

Bring your arm back, take a deep breath, and then very quickly knock away your thick card. When you open your eyes, you'll see that rather than go shooting off across the table, your stunt-egg will have fallen straight down into the now uncovered glass of water. And once you've done this successfully half a dozen times (and managed to keep your eyes open) we both might feel happy with you using a raw egg instead. And don't worry, if the worst comes to the worst, omelets are quite tasty.

If you haven't caused enough mayhem yet:

You can use a larger flat object such as a plastic tray or a placemat and position tubes with eggs on over two, three, or even four glasses. That's right—a multiple high dive! Just make sure all the tubes are lined up accurately above the glasses or the only thing you'll be multiplying is mess … and trouble.

The Sciencey Bit

There is a **force** acting on the egg (or stunt-egg) called **inertia**. This is a force that resists movement—and the greater the **mass** of the object the higher the inertia. The tubes have a lower mass so have lower inertia which means they move to the side when you hit the card, but the high inertia of the egg means it will resist the sideways movement and drop into the glass below instead.

Tip: Try to use a good-sized glass that isn't too delicate.

Tip: Newer nail polishes work best.

Make marbled candleholders

For this project you're going to need to get hold of a film-forming **polymer** dissolved in a volatile organic **solvent**. Stop fretting—it's not as difficult as it sounds—it's just nail polish. Although actually, if it's your mum's favorite color, it might well be *more* difficult than it sounds.

Having convinced your mum that you need her lovely nail polish for a very important demonstration of **surface tension**, you can now get everything set up. First, cover the table with old newspapers or a cloth so you won't spill nail polish on it. Next, get hold of a disposable aluminum container. If you haven't got one, you could use a plastic bowl, but just make sure that if you ever want to use it again, you put a plastic bag inside it so the water—and nail polish—touches this and not the bowl itself.

Now fill the container with water about three-quarters full and get your jar and toothpick close at hand. When you're fully prepared, take your chosen nail polish colors and drip them carefully, from the lowest height you can, into the container. Rather than sink to the bottom, the nail polish is so light it should form a film on top instead. Now quickly use your toothpick to drag across the polish in different directions to create a marbled pattern. When it's done, you need to place your jar into the water—either dipping it in base first, or rolling it on its side over the polish. Just don't hang around or it won't work as well.

You will see that the polish clings to the glass, forming beautiful patterns. Now just place it, top down, on some newspaper until the polish is dry and then you can place a candle at the base making a beautiful "Sorry-I-used-up-all-your-nail-polish" gift for your mum.

If you haven't caused enough mayhem yet:

Try marbling other objects such as napkin rings, plain mugs, glass bottles, blown eggs—just make sure you ask permission first!

The Sciencey Bit

The surface of the water holds together like a skin because of **surface tension**—this is a **force** of **attraction** between all the tiny water **molecules** on the surface of a liquid. As the nail polish is less **dense** than water it doesn't break the surface tension and instead can sit above the water until it is attracted to the surface of the glass jar.

Learn Morse code

YOU WILL NEED:
A FLASHLIGHT, A FRIEND

"No more talking, you two."

"Be quiet, it's bedtime."

"Stop chatting now."

"I don't want to hear a peep out of you."

I know it's weird but there are times when adults don't seem to want you to talk *at all*. Thankfully, they rarely say "No communicating", so these are the perfect moments to crack out the Morse code.

Morse code can be tricky to learn. but once you've got the hang of it, you can pass on messages simply by turning a flashlight on or off. So as long as you can see the light, you can keep chatting.

The code was invented over 175 years ago and uses a combination of dots and dashes—or dits and dahs—to make different letters. A *dit* is short, while a *dah* is three times longer. Every letter is separated by a short pause, and every word is separated by a longer pause.

It's best to learn a letter at a time, starting with the ones you use most often in words, and also by slowing things down until you get more confident. Then why not practice with some important words you know you'll use a lot—"poop", "snot", "butt" for example, before moving onto whole phrases, such as "They think we have stopped talking. Ha, ha."

If you haven't caused enough mayhem yet:

Try using sounds as well as light. You can make the "beeps" yourself or use something that will make a shorter and longer noise—a saucepan lid, for example, can make a long or short note depending on where you hit it.

The Sciencey Bit

Morse code is called a **digital system** because it relies on a series of pulses or light flashes that have just two states—on or off. Talking is an **analog system** because it varies in **pitch**, volume and length of note.

Morse code alphabet

A	.—	J	.———	S	...	1	.————
B	—...	K	—.—	T	—	2	..———
C	—.—.	L	.—..	U	..—	3	...——
D	—..	M	——	V	...—	4	....—
E	.	N	—.	W	.——	5	.....
F	..—.	O	———	X	—..—	6	—....
G	——.	P	.——.	Y	—.——	7	——...
H	....	Q	——.—	Z	——..	8	———..
I	..	R	.—.	0	—————	9	————.

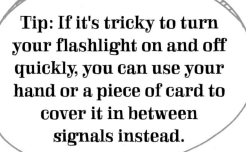

Tip: If it's tricky to turn your flashlight on and off quickly, you can use your hand or a piece of card to cover it in between signals instead.

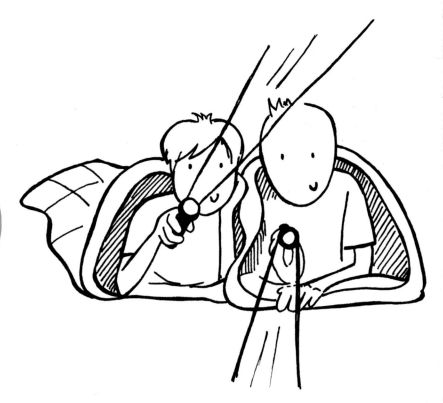

Try some color swap magic

Look on the other page and you'll see a picture— or more accurately, a slightly weird picture. Because yes, I know the sun shouldn't be blue or grass red, and the sky isn't generally yellow. But don't worry, we'll sort that out with a bit of color swap magic.

All you need to do is get a blank sheet of white paper and add a black dot to the middle of it, then put it next to this book. Now you need to stare at this picture—okay, sorry, this *weird* picture— while you slowly count to 30.

Done that? Great! Then immediately afterwards stare at the little black dot on your sheet of paper. After a few seconds you should see the picture appear around the dot but this time—yes, it's not weird, is it? In fact the colors are just as you would expect—yellow sun, yellow and black bee, green lawn. See I told you—color swap magic!

If you haven't caused enough mayhem yet:

Try making your own weird picture swapping colors around—just remember the three pairs you can swap are red and green, blue and yellow and black and white.

The Sciencey Bit

Science hasn't fully explained this phenomenon but we do know that the back of your eye—the **retina**—has millions of tiny **light** sensing **cells** which can see different colors of light. They are hooked up in pairs: white with black, blue with yellow and red with green, and each pair sends information to the brain. Stare at one color for too long and the sensors for that color become saturated causing the brain to interpret it as the opposite color in the pair when it sees the afterimage on the white paper.

YOU WILL NEED: WHITE PAPER, BLACK PEN, YOUR EYES

Go fishing

Fishing usually involves long rods, reels, nets and oodles of patience. Luckily for you, this game requires none of the above—which is good because that last one is almost impossible for children to locate.

Begin by drawing a fish shape on your piece of card—when you're happy with how it looks, cut out this template and use it to trace six fish shapes onto each different piece of colored paper. When you have cut these out, draw a number of spots on each set of fish from one to six. Now slip a paperclip on the end of each and spread the fish face down on the ground so you can't see how many spots each one has.

To make your short rods, take a stick or pole (one per player) and tie string to the end. Cut this so it is about 20 inches long and then attach your **magnet** to the other end using the tape.

Now you are ready to play the game. Sit a little way from the fishes and take it in turns to dangle your rod over them until you manage to pick one up with the magnet and put it in your caught pile. Be careful though—pick up more than one and you will be deemed a "greedy fisherman" and have to throw them back into the sea, missing your turn.

At the end of the game, you can count the number of spots on the back of your fishes, with the highest number declared the winner, presented with the golden rod*, paraded through the town** and awarded the gift of everlasting patience***.

*optional

**optional

***impossible

If you haven't caused enough mayhem yet:

Why limit yourself to fishes? You can make any sort of template to cut out shapes for your game so why not try catching dinosaurs, ghosts or flowers instead?

The Sciencey Bit

Paperclips are made from steel which includes **iron**—a metal that is **attracted** to **magnets.** This means the magnet is attracted to the paperclip, which lifts the paper fish off the ground.

Tip: If you don't have any paperclips, you can add a couple of staples to the end of each fish instead.

Make your own skeleton

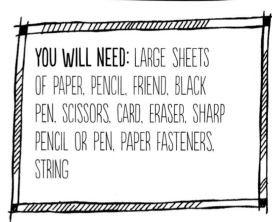

YOU WILL NEED: LARGE SHEETS OF PAPER, PENCIL, FRIEND, BLACK PEN, SCISSORS, CARD, ERASER, SHARP PENCIL OR PEN, PAPER FASTENERS, STRING

Skeletons are supposed to be scary—which is a very odd idea. After all, the scary thing would be life without skeletons—we'd all look like deflated balloons lying on the ground. Now that *is* terrifying.

So why not celebrate your own skeleton by recreating it?

To add accuracy to this model, you'll need large sheets of paper and some help. Lie on the ground with your head on a sheet of paper and ask your friend to trace around your head and neck. Then do this with one upper and lower arm (including your hand) and one upper and lower leg.

When they are tracing around your main body, feel down your sides to locate your ribs and then ask your friend to make a mark where each one ends. Next, lie with your bottom on some paper and get your friend to trace around and also mark where your hipbones (pelvis) sit. Finally, place your feet on a piece of paper and draw round these yourself (after all, no friend should have to deal with your smelly feet).

Now you have your outlines you can start forming the bones. Use the guide opposite to help you draw each one in pencil and mark where the holes need to go, then cut them out (add an extra sheet below your leg, arm and hand so you cut out a pair of each). Use a black pen to add teeth as well as eye sockets and nose holes to your skull.

To strengthen the structure, stick the skull, central section and pelvis to pieces of card (old cereal boxes work well) and cut these out.

Now it's time to pull yourself together. Put each hole guide over an eraser and push through a sharp pencil or a pen to make each one. Now insert a paper fastener into each joining pair and open it at the back to secure it. Finally, if you want to dangle your skeleton, add a hole to the top of the skull and thread it with string before attaching it to a hook or nail on a wall or doorframe.

If you haven't caused enough mayhem yet:

You could try making skeletons of your whole family—that'll freak them out!

The skeleton

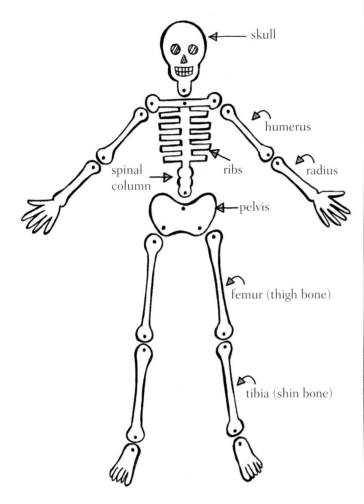

- skull
- humerus
- ribs
- radius
- spinal column
- pelvis
- femur (thigh bone)
- tibia (shin bone)

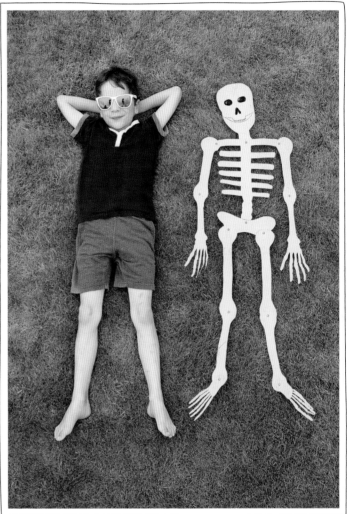

 **The Sciencey Bit**

The human **skeleton** is made of more than 200 individual bones and helps protect some of our vital body parts like the brain and the heart. They also give us support so we can stand upright and allow us to move about because **muscles** are attached to bones. Did you know that red and white blood **cells** are also made inside our bones?

Tip: If you don't have large sheets of paper, just tape some smaller sheets together when needed.

Hang a pendulum

"We'll go to the park in a minute."

"Yes, you can have a cookie, just give me a second."

"Five minutes, and we'll go for ice cream—I promise!"

If these phrases prove anything, it's that parents are *awful* at telling the time. Thankfully, it is very easy to bring a bit of accuracy into their life—you just need to hang a pendulum.

To do this, push a thumb tack into the top of a wooden doorframe (you'll need to ask your parents' permission, and as it'll be hard for you to reach, you could probably get them to help too), then tie on a piece of string, just over one metre in length, to the pin.

Finally, tie a weight on the end—a key or a washer works well as these already have a hole to thread the string through. Do this carefully, checking against a tape measure to try and make the final length of the string 39 inches long, then pull the weight to one side and let your pendulum swing.

What you will find is that each separate swing of the key or washer takes one second. It should keep swinging for a while, allowing you to time your parents. This means whenever they say "a couple of minutes" or "half a minute" you can let them know, very loudly, exactly when that time arrives—especially if ice cream is involved!

If you haven't caused enough mayhem yet:

Try making a half-second pendulum. This will need a shorter length of string so you could hang your pendulum from a pole stretched between the back of two chairs. You'll need a stopwatch to work out which length of string gives exactly the right measure of time.

The Sciencey Bit

A pendulum swings back and forth because it is being pulled by **gravity**. It will eventually stop swinging only because of resistance between the air and the weight and **friction** between the string and the thumb tack, both of which will gradually slow it down.

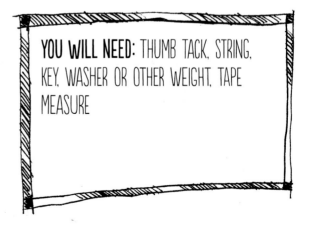

YOU WILL NEED: THUMB TACK, STRING, KEY, WASHER OR OTHER WEIGHT, TAPE MEASURE

Tip: To avoid pushing pins into wood, you could try hanging your pendulum from curtain rails instead and swing it from side to side.

Build a periscope

There are many times when you may need to hide—when you've *mistakenly* eaten an entire chocolate cake, or *accidentally* broken your brother's favorite game, or *unintentionally* destroyed your sister's school project. But the problem with hiding is it's very hard to see if your family are about to find you. Do you know what you need? A periscope.

Yes, this is the perfect way to check out what is going on while still cowering behind the sofa.

Begin by taking your card and folding it over to form a right-angled triangle. Line this up against the side of one of your cartons so you can use it as a template to draw a 45-degree line (see diagram). Do the same on the other side to make sure the marks are lined up, and then repeat this on the base of your other carton.

Now measure and mark $1/3$ inch in from either end of these lines—this middle area is the line that needs cutting. Make an incision hole for your scissors by taking a safety pin, pin or thumb tack and pushing this through again and again along the start of the line, with each hole as near to the next as possible. You should then be able to push through the end of your scissors on this section of perforated line and cut the rest of the slit. You can also use the scissors to widen this slit if needed.

Now use this same method to cut a square window opposite each slit in both cartons (see diagram) before opening up the tops of each carton and slipping one onto the other. Just make sure that the windows sit on opposite sides.

Finally, add your mirrors by slotting them through the slits. If the mirrors are narrower than the gap between the slits, glue them onto a piece of card a little wider than the gap and slot this through.

And now you are ready to use your periscope—which is good because we both know you are accidentally about to eat another whole chocolate cake, don't we?

If you haven't caused enough mayhem yet:

If you cut the base out of more juice or milk cartons you can use these to extend the middle section of your periscope and make it even longer. Very handy when you need to see what's going on a whole floor above where you are hiding.

The Sciencey Bit

Light travels in straight lines, so it enters the top window, but then by positioning your mirrors at exactly 45 degrees it **reflects** off the top mirror and down the tube to the mirror on the bottom where it again reflects straight to your eye through the lower viewing hole.

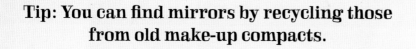

Tip: You can find mirrors by recycling those from old make-up compacts.

Making the periscope

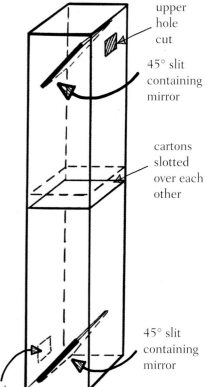

upper hole cut

45° slit containing mirror

cartons slotted over each other

45° slit containing mirror

lower viewing hole

Tip: You can use stickers, acrylic paint, pens or ribbon to decorate your finished perfume jar.

YOU WILL NEED:
SCENTED LEAVES OR FLOWERS, SMALL, CLEAN, EMPTY, GLASS JARS, OLD SPOON, WATER, PEN OR PENCIL, PAPER OR NOTEBOOK, BOILING WATER (AND ADULT), SIEVE, INSPIRATION

Make perfume

Yes, you too can join the legions of celebrities who launch their own perfumes. Except yours will be better. Obviously.

First, you're going to need to find some fragrances and the garden is the perfect place to look. Some good flowers to try are roses, honeysuckle, lavender, jasmine, and violets, but any scented leaves or flowers can be used so spend a bit of time smelling what's on offer. Also, remember to ask an adult's permission before you start stripping petals from their favorite plants.

When you've chosen your raw materials, put a couple of tablespoonfuls in a jar and give them a good pounding with your spoon—this bruises the petals or leaves and helps to release more scent. Now cover them with a little drop of water, give it a stir, and take a good sniff. You may have hit on the perfect combination straight away, but equally you can have a go at experimenting—increasing the amount of some ingredients if their smell is too subtle to be noticeable or adding new ingredients to bring in new scents.

To keep track of your work, you can create different combinations in separate jars, noting down which ingredients you use and how much of each. This way you can pinpoint which combinations work best and also recreate it another time.

When you're happy with your recipe, fill a jar two-thirds full with your chosen ingredients, give them a good pounding and then ask an adult to pour some boiling water on top. Leave it to sit for about an hour so the scent is released and then carefully strain through a sieve into a jug then pour it back into a clean jar and replace the lid. This 'perfume' should now keep for a couple of weeks.

To try it out, dot a bit of the mixture on your wrists, and give it a sniff. Oh, and remember to give it a suitably "perfumy" name: maybe Eau de Back Garden, or Catastrophe for Men!

If you haven't caused enough mayhem yet:

As well as floral scents, why not try creating some fruity or spicy or even woody fragrances too? Which scents are strongest and last longest?

The Sciencey Bit

Plants produce scents, not to help the perfume industry, but to attract **pollinating** creatures and insects. These scent **molecules** are known as **volatile** because they **evaporate** into a **gas** when they are released by the plant. This also means that they are released into the air when you place the perfume on your wrists, allowing you to smell its beautiful fragrance.

Design a beach mandala

A mandala is traditionally a way of representing the cosmos. Yes, that's right—you're going to create the entire universe!

Thankfully, this is a lot less tricky than it sounds and usually involves forming patterns in circles.

What you do need to do is find lots of different materials for each ring of your mandala. Begin by collecting a few buckets of shells and pebbles from the beach and then sort these into separate piles—you can start with "rocks" on one side and "shells" on the other. Now keep sorting these even further into different groups, maybe by color and then size for stones and perhaps by size and shape for shells—by doing this you are setting up a **classification system**.

If you take along a spotters' guide, you'll even be able to say which animal a shell belonged to or which sort of rock each pebble is. As you do this, you can use your finger in the sand to write the name of each one beneath its pile.

When you have all your raw materials classified, it's time to start creating your mandala. The nearer the center you are, the smaller the circles, so the fewer stones or shells you need. Just remember, when you move farther out you're going to need a lot of whichever color, shape, or type of material you've selected, so check your classification system to see which shell or pebble is most common.

And there's no need to simply stick to concentric circles. You can bring in all sorts of different designs to radiate from the center, maybe some petal shapes or diamonds … or … well anything really. It is your cosmos after all.

This can be a good activity to do in a group—and no, this does not involve you just sitting around demanding people "Get me 35 whelk shells—NOW!" That's no way to run the cosmos, you know.

And don't forget at the end to drag over the adults and say "You may be interested to know, while you have been doing nothing more than lazing about on your towel, we have created a graphical representation of the cosmos. And if that isn't worth an ice cream I don't know what is!"

If you haven't caused enough mayhem yet:

Try adding seaweed to your design—and see how many different types you can identify. You can also make a **classification key** starting with "Is it an animal, plant or rock?" And working down to each individual **species** or type.

 The Sciencey Bit

Although the pattern of your mandala is nice to look at, this activity is all about sorting things into groups. In science you can group things by a **classification system**. This means you start off with a broad grouping like "seaweed" or "seashells", but as you look carefully and spot differences you can divide this down further, maybe into shells with bivalves (two hinged shells) or single shells.

YOU WILL NEED: A BEACH, STONES AND SHELLS, A FIELD GUIDE/SPOTTERS' GUIDE

 Tip: Larger stones or shells are best saved for outer layers as you will need fewer of them to make the circle.

Crystallize flowers

These brilliant cake decorations will make everyone gasp and say "Wow—those flowers look *almost* real." This gives you two equally appealing options: either look smug and take the praise or roll your eyes, sigh and say "That's because they *are* real … obviously."

But before you get to this happy stage you need to prepare your crystallizing station. Place your separated egg white in a bowl, add a tablespoon of water and whisk this gently for a minute with a fork to break it up a bit. Then put out a container with some superfine sugar in it, a plate or shallow bowl, and a piece of greaseproof paper.

Next, go and pick a few edible flowers, ensuring that they are clean and dry—rose and sunflower **petals** work well, as do violas, primroses, cornflowers, pansies, pelargoniums, and borage. Just make sure you get an adult to double-check you have the right flowers and if in doubt, don't use them (some flowers are actually poisonous!).

Now grip the flower by the stalk or, if you are working on petals, carefully hold one end between your finger and thumb. Use a small clean paintbrush to coat all the surfaces—front and back—in a thin layer of the egg white mixture.

Take a pinch of superfine sugar and sprinkle this over the flower until it's covered, then lay it, very carefully, face down on the greaseproof paper. In an hour or two it will have become quite solid and you can check for any uncoated bits you need to redo.

After you've let the flowers dry for a day you can carefully remove any stalks or sepals that are left on (see diagram) and then use the flowers to decorate cakes and puddings. And if you have any left over, you can store them for a few weeks between sheets of greaseproof paper in an airtight container. After all, it's always good to have a few "so-impressive-you'll-gasp" decorations to hand—just in case you need that lovely smug feeling.

If you haven't caused enough mayhem yet:

Why not try crystallizing some edible leaves? Mint leaves are ideal and look great alongside the flowers.

The Sciencey Bit

Bacteria and **fungi** would usually cause these flowers to **decay** and rot but to stay alive bacteria and fungi need **water**. Sugar is a **hygroscopic** substance, which means it draws water **molecules** from the flowers by **absorption**, stopping the bacteria and fungi doing their work.

Parts of a flower

anther

stigma

stamen

petal

sepal

stem

Be a rock sleuth

There is a fascinating world to discover underneath your feet. Eughh—no, not your verrucae. I mean the rocks that make up the earth. Now please put your grubby feet away— we've got some **geology** to get on with.

First, to become a rock sleuth you'll need to gather some rocks. Have a look around your garden, the local park, woods, on a walk into town—they really are everywhere. What you want to do is find as many different ones as possible.

Now it's time for some detective work. The first test you can do is to find out how hard they are to break—and what happens when, or if, they do. For this you'll need a hammer (and some adult supervision), safety goggles and an old towel to wrap your rock in. Put the towel-wrapped stone on the ground outside, smack it with the hammer and then open it up to see what happened. If it's broken into rough or jagged pieces it has fractured—other rocks might cleave, which is when they break into smooth flat pieces.

You can also use the Mohs scale of hardness (see opposite). Whether or not different things can mark your stone will show you how hard it is, which is also a way of learning what minerals are in it. Use a copper coin and an **iron** nail to test against the scale and try and give your mystery rock a score.

Finally, you can decide whether your rock is an igneous, sedimentary or metamorphic rock with the following clues:

Igneous clues: has lots of small holes, is very hard to smash, and has light and dark speckles or large visible crystals—these rocks are made when melted minerals cool and harden and come from volcanoes!

Sedimentary clues: has layers, breaks apart easily, highly porous so will **absorb** water (to test this, weigh it, soak it in water for two minutes, and weigh again). These rocks are made when lots of small rock **particles** are pressed together.

Metamorphic clues: has swirled patterns, is hard to smash and has an all over shininess (rather than speckles). These are made when rocks buried deep in the earth are squeezed together and heated for millions of years.

If you haven't caused enough mayhem yet:

If you can find an old bathroom tile you can use the back (non shiny) side to perform a streak test. For this you drag the rock along the surface and see what color streak it leaves behind, which you can then check against a guide to help you identify your mystery rock.

Mohs Hardness Scale

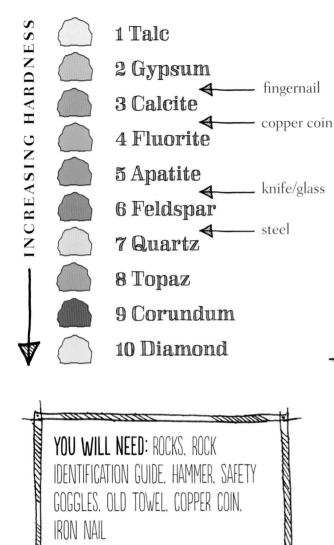

INCREASING HARDNESS

1 Talc
2 Gypsum
3 Calcite — fingernail
4 Fluorite — copper coin
5 Apatite
6 Feldspar — knife/glass
7 Quartz — steel
8 Topaz
9 Corundum
10 Diamond

YOU WILL NEED: ROCKS, ROCK IDENTIFICATION GUIDE, HAMMER, SAFETY GOGGLES, OLD TOWEL, COPPER COIN, IRON NAIL

Tip: Try putting a few drops of vinegar on your rock and seeing if it fizzes—vinegar is an acid which dissolves calcium carbonate, a major ingredient of limestone, which is a type of sedimentary rock.

The Sciencey Bit

Geology is the study of the earth and its rocks and minerals and you have been performing some of the key tests done by geologists.

Make rainbow paper bookmarks

Rainbows are well known for brightening up the sky or pinpointing Leprechaun gold hoards. But you might be surprised to find that they also make very useful bookmarks. Yes, really, they do.

First, use a pencil and ruler to measure out a long rectangular shape on a piece of black card, which you can then cut out to form your bookmark base.

Protect your bowl with plastic wrap or aluminum foil and then half fill it with water. Place your card on the bottom and add a couple of drops of clear nail polish to the surface. Do this as close to the water as possible and watch as it spreads out. Now pull your card up through the surface, tilting it slightly, so it is coated in a thin layer of **floating** polish.

Put your card on some old newspaper to dry, propping it up a little on one side so the extra water can drain off. When it's fully dry, you can flatten it out by placing it between some heavy books for half an hour.

Finally, punch a hole through the top of the bookmark and thread through thin pieces of ribbon in rainbow colors which you can tie together to hold them in place.

Now you have a beautiful bookmark which you can tilt in the light to see rainbows appear. It is also a perfect gift—especially for any Leprechauns wanting to mark their place in *How to Hide your Gold: Volume 1*.

If you haven't caused enough mayhem yet:

Why not cut out circles of black card and make rainbow coasters? If you coat these in clear contact paper it will protect the card from spillages.

The Sciencey Bit

The film of nail polish is only a few hundred **nanometers** thick (a nanometer is a billionth of a metre). As the film is slightly thicker in some places than others it **refracts** or bends the **light** more in those areas, splitting it up so you see the different colors which make up white light.

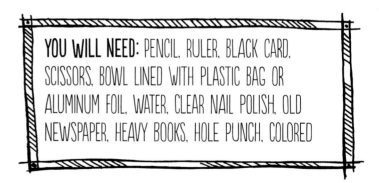

YOU WILL NEED: PENCIL, RULER, BLACK CARD, SCISSORS, BOWL LINED WITH PLASTIC BAG OR ALUMINUM FOIL, WATER, CLEAR NAIL POLISH, OLD NEWSPAPER, HEAVY BOOKS, HOLE PUNCH, COLORED

Create a compass

It's important to understand which direction you're facing—after all, how else will you know when birds are flying south, if that wicked witch is really from the west, or where to watch for Santa's arrival from the North Pole?

Luckily for you, it's pretty easy to create your own compass. All you need to do is draw around a small plate or bowl to make a circle. Cut this out, then fold it into halves, quarters, then eighths and unfold. You can now mark the points top and bottom as North and South, those on the left and right as West and East and those between as North East, South East, South West and North West.

Next, fill a glass about three-quarters full with water, place it on the middle of your paper circle and **float** your bottle cap upside down on top.

Finally, take a strong **magnet** and stroke your needle in the same direction, from the blunt to the pointy end, 20 times before dropping this carefully across the top of the bottle cap.

Soon you will see the cap move and when it stops, the sharp end of the needle will be pointing North and you can twist your paper dial around beneath the bowl until this matches up … and then sit and wait for Santa to arrive.

If you haven't caused enough mayhem yet:

Try integrating your dial and disk by getting a larger plastic bottle top and marking N (North), S (South), E (East) and W (West) on it with a permanent marker.

The Sciencey Bit

The needle is made of **particles** of **iron** that have **charges** which are usually pointing in many different directions cancelling each other out. When you stroke the needle with the **magnet** this causes the charges to all point the same way, making the needle behave like a magnet itself. Inside the earth there's so much iron that it acts like a giant magnet with a **pole** at each end. Now it is magnetized the needle wants to line up in that direction too, with its south pole pointing towards magnetic north because opposite poles attract. **Floating** on water allows the needle to move freely and do just that.

Mix bath salt gift jars

Adults often like to take a long soak in a bath. I've no idea why. Anyone would think they worked hard and needed to relax. As if! Still, it's wise to keep in their good books, so why not gift them some relaxing bath salts?

Fill your jar halfway up with Epsom salts and then to the top with sea salt so you know you have the right amount. Now tip this into a bowl and mix together, along with half a teaspoon of glycerin if you have it (this helps the colors and oils spread more evenly, but isn't necessary) and a few drops of your essential oil (which is … essential). Now divide your mix into seven equal measures and into each add one or two drops of food coloring to make your different rainbow colors.

Now begin to spoon your colored salts into your jar, starting at violet and working backwards, until it's full of your rainbow bath salts.

Finally, make a label on your piece of card. Make sure it says what sort of oil you have used and that these are bath salts—not a tasty snack. Also, you could write that half a cup in a warm bath *may* help people relax (even if they haven't really done very much), and can stop them getting wrinkly fingers. Punch a hole in one corner of this informative label, thread through your ribbon, and tie it to the neck of your jar.

If you haven't caused enough mayhem yet:

Try a variety of different colors and essential oils to find your perfect combination.

The Sciencey Bit

Osmosis is the movement of water through a **membrane** (such as your skin) to achieve a balance. Your body contains water and salt, whereas an ordinary bath contains mainly water and very little salt. Therefore, water passes through your skin in an effort to balance the concentration of water between you and your bath. This excess water causes 'pruning' (your fingers and toes wrinkle). Adding bath salts to the water causes a more equal balance of salt and water in both you and in the bath, so less water enters your skin and less wrinkling occurs.

YOU WILL NEED: SEA SALT, EPSOM SALT, ESSENTIAL OIL, FOOD COLORING, SPOON, JAR, CARD, PEN, HOLE PUNCH, RIBBON, GLYCERIN (OPTIONAL)

Grow magic roots

Plants are pretty good at making lots more of themselves—and not just from **seeds**. In fact, they can quite often grow new **roots** from a tiny snipped off stem, which is a bit like growing a new pair of feet out of a tiny finger—except *much* less weird.

Usually this clever work goes on under the ground but some plants will let you peek at what they do because they'll even produce new roots in water. If you want to watch the magic in action, first you'll need to find some suitable plants to snip. Many herbs are good to try such as rosemary, mint, sage, tarragon, lemon balm, thyme and oregano.

Cut about 6 inches from the end of a plant stem without flowers, strip off any leaves on the bottom half and then place this "cutting" in a small container of water on a sunny windowsill. It's good to find a container with a slightly narrower top as this will help hold the stem upright.

You should change the water every few days so it stays fresh—you can use tap water but rainwater is even better as it contains fewer chemicals and more **nutrients**.

Amazingly within 2–6 weeks (depending on the plant) you should start to see roots begin to grow out of the base of the stem. You can keep them this way, harvesting their new leaves if they are herbs, or else carefully plant them in small pots of soil and try growing them on.

If you haven't caused enough mayhem yet:

Why not try taking other leaf, root or stem cuttings from different plants and see how many you can grow? Which do better—those in water or those in potting soil? Why do you think this is?

The Sciencey Bit

Roots which grow from somewhere other than the original **seed** are called **adventitious roots**. When plants are grown from other parts of a plant such as a stem, leaf or root cutting this is called **vegetative propagation** and results in the plants being **genetically** identical to their parents— or **clones**.

Tip: Unlike leaves, roots prefer to grow away from light so you could use darker glass containers or even wrap some paper around the container.

Try shaving foam marbling

This technique will give you the effect of marble without the need to actually mine rocks or handle dynamite—although it might be just as messy.

First, cover your surface with a wipeable cloth or old newspapers. Next take your tray and squirt in some shaving foam which you can then spread with a ruler or smooth it with a spoon to make sure it covers the entire surface.

Now get your paints and start squirting or dribbling them on top of the foam. Try to make sure there is a good balance of paint around so no area is left too bare. Using two or three different colors works well.

Take a toothpick or skewer and begin to create your marbling effect. You can drag the paint in lines or waves or even add swirls—whatever looks best. Just make sure you tweak all the areas so no big blobs of paint are left untouched.

Next, lay your paper on the surface and press down gently on it, all over, so that every piece is in contact with the marble mixture beneath. When you're happy, peel off the paper and lay it down on your covered surface.

Now this is important: DO NOT PANIC! It will look like a gloopy mess, but just wait for a minute then take your ruler, place it on its edge at the top of the sheet of paper and pull it across the page. Amazingly, when the "gloop" is removed, you will be left with a beautiful marbled pattern beneath which just needs to be left to dry.

Your kitchen table, however, will still look like a gloopy mess. This is important: DO PANIC—and make sure you clear it up, or you will never get to marble paper again.

If you haven't caused enough mayhem yet:

Try using card instead of paper—which works best? What about comparing different kinds of paint—liquid watercolors, acrylic paints, poster paints, even food coloring—which were most effective?

The Sciencey Bit

The water-based paint doesn't mix with the **hydrophobic** (water hating) ends of the **molecules** in the shaving foam so instead sits on the surface. However, the water-based paint soaks into the paper because this is made of **cellulose** which is **hydrophilic** (water loving) meaning the water in the paint is attracted to it.

YOU WILL NEED: WIPEABLE TABLECLOTH OR OLD NEWSPAPERS, TRAY WITH SIDES (BAKING TRAYS WORK WELL), SHAVING FOAM, RULER, PAINTS, TOOTHPICK OR WOODEN SKEWER, PAPER OR CARD

Tip: You can use marbled paper to wrap books or make artwork or gift bags (see page 66), and use card for gift tags or bookmarks.

Tip: If you place your toothpicks on a shallow plate, this will give you a smooth surface as well as keeping the table dry.

Create a magic star

So toothpicks can be used either to dislodge bits of old food stuck between your teeth, or to create a magical star. Come on, let's give those toothpicks a better life!

Begin by snapping five toothpicks halfway along so they are hinged in the middle—yes, okay, as better lives go, this doesn't sound a promising start, but it'll get better, I promise. Now place the broken toothpicks on a plate so that the middle parts are all arranged around the center and the points around the outside.

Finally, you need to place a drop or two of water in the center of your toothpicks. If you don't have a pipette or eyedropper, you can do this by placing a straw in some water, putting your thumb over the end and lifting it out. With no **air pressure** pushing down from above, the air pressure around the end of the straw will keep the water in the straw until you hold it over the center of the picks and remove your thumb.

Now watch as the toothpicks begin to move and rearrange themselves into a beautiful star. Marvel at this, show your friends … and then throw the toothpicks in the bin. What's the matter with that? It's still got to be better than dislodging half-chewed bits of meat from between your molars—surely?

If you haven't caused enough mayhem yet:

Try using more sticks—does this work as well as five? Or how about trying with something larger and thicker like kebab skewers?

The Sciencey Bit

The toothpicks are made of dry wood and when the water is placed in the middle of the broken sticks, they begin **absorbing** the water, causing the bent wood **fibers** to expand and straighten out. As the toothpicks straighten and push against each other, the inside of the star opens up.

YOU WILL NEED: TOOTHPICKS, PLATE, WATER, CUP OR GLASS, STRAW, PIPETTE OR EYEDROPPER (OPTIONAL)

Launch some raisin submarines

"May I have a fizzy drink?"

If the answer to this question tends to be "No" then might I suggest a different tactic?

"May I have a fizzy drink … so I can conduct a science experiment?"

I'm guessing that'll get a very different response. Even better, you'll need to grab a few raisins too. Yep, snacks and a drink—now we're talking.

Of course, you are actually going to have to do the experiment first, but thankfully it's an absolute doddle.

First, fill a glass three-quarters full with a fizzy drink and then drop in a few raisins. As you watch you should see the raisins being surrounded by lots of bubbles until, whoosh—up they **float** to the surface like a submarine.

And now, also like a submarine, they float on the surface for a bit before diving back to the base.

They will continue like this until there are fewer bubbles when the poor raisins will be stranded at the bottom of your glass. At this point, the only thing you can do is scoop them out, gobble them up, drink the lemonade, and start again.

If you haven't caused enough mayhem yet:

Why not compete with a friend? Both of you can select a raisin, drop it into the glass and see which makes it to the top first—or maybe the first to surface and sink three times.

The Sciencey Bit

Fizzy drinks are also called **carbonated** drinks because they are full of tiny bubbles of a **gas— carbon dioxide**—and as these are less **dense** than water they rise to the top. When you drop a raisin in, bubbles stick to its surface until there are so many they are able to **float** the raisin to the top, where some bubbles burst and the raisin drops down again … until more bubbles stick to its surface and the process begins again.

YOU WILL NEED: GLASS, FIZZY DRINK, RAISINS

Tip: Try to avoid dark-colored fizzy drinks, as these will make it harder to see the raisins.

Resources

USA

Target
Accessible household supplies, from essential craft kits to stationery and games.
target.com | 1-800-591-3869

Michaels kids
Creative screen-free supplies for crafty afternoons, including paints, card, glue, balloons and yarn.
michaelskids.com | 1-800-642-4235

Blitsy
Wide range of craft supplies from painting tools, markers, accessories and more.
blitsy.com | (855) 813-3429

Kmart
Craft supplies including paints, crayons, yarn and paper; garden supplies including pots, seeds and potting soil; kitchen supplies including muffin pans and baking sheets.
kmart.com

Joann Stores
Craft supplies including paints, card, paper and fabrics, yarn, string, ribbon and sewing thread.
joann.com

Glossary of terms

Absorb—take something in.

Acid—a kind of chemical that reacts with **alkalis**. Strong acids can **corrode** metals and are very dangerous. Weaker acids, such as lemon juice, add a sour, tangy taste to drink and food (very important for fans of lemonade).

Air molecules—a mixture of **molecules** of different **gases** that make up the air around us.

Air pressure—the **force** of air pressing down on an area.

Alkali—a kind of chemical that reacts with **acids**. Strong alkalis can cause burns and are very dangerous. Weaker alkalis, such as baking soda, taste bitter and have a soapy, slippery feel when made into a paste. Another word for an alkali is a base.

Atom—one of the very tiny **particles** that everything is made from.

Attract—to pull something closer.

Bacteria—single-celled living things that are found everywhere but can only be seen using a **microscope**. They can be dangerous, such as those that cause infection, or useful, like those that live in our guts (lucky things!) and help keep us healthy.

Carbon dioxide—a **gas** made up of one carbon and two oxygen **atoms**. It is breathed out by humans and animals and, very importantly, is the gas used to make fizzy (carbonated) drinks.

Cell—a small unit of a living thing that contains **genes**.

Cellulose—the main ingredient in plant **cell** walls and of vegetable **fibers** such as cotton. It cannot **dissolve** in water.

Characteristic—a distinctive quality held by something or someone, such as eye color or the ability to wiggle your ears.

Charges—these can be either negative or positive. Opposite charges (i.e. positive and negative) **attract** each other.

Chemical reaction—when one or more new substances are made when mixing two or more things together.

Chlorophyll—a green **pigment** in plants which **absorbs energy** from sunlight as part of **photosynthesis**.

Condensation—the change of state of a substance from a **gas** to a **liquid**.

Corrode—to slowly break apart and destroy through a **chemical reaction**.

Data—facts and statistics collected together.

Decay—the breaking down of something into smaller parts (**molecules**) by the work of **bacteria** or **fungi**.

Density—how heavy something is for its size.

Diameter—the length of a straight line going through the center of a circle from the edge of one side to the other.

Dissolve—to break up something (a **solute**) into small parts and mix them evenly into a **liquid** (the **solvent**) to form a **solution**.

DNA—short for Deoxyribonucleic acid (try saying that with a mouthful of marshmallows) which is a large, spiral-shaped **molecule** that carries **genetic** information including that which determines **characteristics** such as eye or hair color.

Electron—a tiny **particle** with a negative electric **charge**.

Energy—the ability to do work. Energy exists in different forms such as heat, **kinetic energy**, light, **potential energy**, sound, chemical energy and electrical energy and can be transformed from one form to another.

Evaporate—to change from a **liquid** into a **gas**.

Experiment—a test or series of tests carried out in order to collect **data** to support or disprove your potentially brilliant idea or theory.

Fiber—a long, fine thread or strand.

Float—being supported on a **liquid** or **gas**.

Force—something that creates a push, pull or twist on an object to make it move, change shape or change direction.

Friction—the **force** that acts in the opposite direction to movement and tends to slow down moving objects, usually when two or more objects rub together.

Fungi—living things such as mushrooms and molds that survive by breaking down and **absorbing** the natural things on which they grow.

Gas—a substance that doesn't have a fixed shape or volume and can fill the space it is in. Also useful for burping.

Gene—found in **cells** and made of **DNA**. **Genes** are inherited instructions for **characteristics** passed from one generation to another.

Genetics—instructions for **characteristics** that you inherit from your parents.

Germination—when a **seed** starts to grow into a plant.

Gravity—the **force** that **attracts** objects to the center of the Earth and prevents us from floating away (phew!).

Hydrophilic—something that **attracts** water.

Hydrophobic—something that repels or pushes away water (rather like you at bath time).

Ice—**solid** state of water (and very useful for keeping your lemonade cool).

Iron—a **magnetic** metal.

Kinetic energy—movement **energy**.

Lift—the upwards **force** that helps things fly.

Light—a form of **energy** which we can see and which travels in straight lines.

Liquid—something that has a certain volume but no fixed shape and can be poured (like lemonade).

Magnet—a piece of **iron** that **attracts** certain other metals.

Mass—the amount of matter there is in an object measured in grams and kilograms.

Microscope—a device that creates a much larger view of very small objects so that they can be seen clearly.

Molecule—tiny particles made up of two or more **atoms** bonded together.

Muscles—bundles of **fibers** in a human or animal that can contract and relax to make parts of the body move.

Nectar—the sweet watery **liquid** in flowers that butterflies and other pollinating insects feed on.

Nerve—a thin **fiber** that sends messages to and from your brain.

Nutrients—the substances **absorbed** by the **roots** of plants or consumed by animals which are needed by both to live and grow.

Osmosis—when water passes through a **semi-permeable membrane** to make the concentrations on either side equal.

Particle—a tiny piece of material so small you cannot see it with the naked eye.

Petal—one of the modified, often brightly colored, leaves around the center of a flower.

Pitch—how low or high a sound is. This is determined by the rate of vibration caused by the sound.

Photosynthesis—the way plants take **energy** from sunlight and use it to convert **carbon dioxide** and water into sugar and release oxygen into the air.

Pigment—substance in things that gives color.

Pollen—tiny grains produced by the male part of a flower that are the male reproductive **cells**.

Pollination—the delivery of **pollen** to the female part of a flower where it can combine with a female reproductive **cell** to achieve fertilisation and produce a **seed**.

Polymer—a long chain of **molecules**.

Potential energy—**energy** that is stored.

Reflect—the way **light** or sound bounces off a surface.

Refract—to bend **light** so it changes direction.

Reinforce—to add something to a structure to make it stronger.

Retina—the back of the eye where light sensitive **cells** receive images and send signals via **nerves** to the brain about what is seen.

Roots—the part of a plant that grows downward, holds the plant in place and **absorbs** water and **nutrients** from the soil.

Seasons—each of the four divisions of the year (spring, summer, fall and winter) marked by particular temperatures and daylight hours, resulting from the earth's changing position to the sun. Also useful when deciding whether to wear woolly hats or flip flops.

Seed—something produced by a fertilized plant that contains a baby plant, called an embryo, wrapped in a seed coat (**testa**) that will start into growth (**germination**) when the conditions are right.

Semi-permeable membrane—the thin, flexible layer surrounding **cells** which will let some, but not all substances, pass through it.

Shadow—the dark area where **light** rays could have fallen if not for an object in the way.

Skeleton—the firm structure of a living thing that in humans is made of bone and supports our body as well as protecting our vital body parts. See—not scary at all!

Solid—a substance that keeps its shape rather than spreading out like a **liquid** or a **gas**.

Solute—the substance that **dissolves** in a **liquid** to form a **solution**.

Solution—two or more substances, including a **solute** and a **solvent**, evenly mixed together.

Solvent—a substance that **dissolves** a **solute** to make a **solution**.

Species—a group of animals or plants that are similar and can produce fertile young animals or plants together.

Spectrum—a series of colored bands of **light** that can be seen when white light is split such as a rainbow.

Surface area—the measure of how much outside area a **solid** object has.

Surface tension—a **force** that **attracts** water **molecules** to one another and that holds the surface of water together like a skin.

Suspension—a mixture in which small **particles** of a substance are dispersed throughout a **gas** or a **liquid** but are likely to settle to the bottom over time.

Translucent—something that lets **light** pass through but scatters it so that detailed shapes on the opposite side cannot be clearly seen.

Transparent—something that allows light to pass through it so that objects behind can be distinctly seen. Also *not* very useful when playing hide and seek.

Categories

Best for biology

Fun for physics

Cool for chemistry

Good for younger children (age 3–7)

Good for older children (age 7+)

Quick projects

Perfect for presents

Crafty projects

Foodie projects

No mess activities

In the garden

Index

Acknowledgements

I must give credit to Kyle Cathie—the power behind Kyle Books—for suggesting that I tackle science for the latest addition to the "101 Things" series. Thankfully what could have been a daunting task was made far easier by the wise counsel of Liz Turner—a science teacher extraordinaire and my wonderful fact checker for all the projects.

As always, the words are just the start. The book comes to life thanks to the beautiful photographs taken by the talented Kate Whitaker, the eye-catching book design from Louise Leffler and brilliant illustrations drawn by Sarah Leuzzi. And of course the whole process has been expertly shepherded from conception to publication by my wonderful editor Tara O'Sullivan.

Thanks also to my junior scientists who modelled for the shots in the book: Chloe, Tom, Emily, Poppy, Michael, Enrico, Carlos, Iris Z, Isla, Katie, Joshua, Ruby, Jake, Amelia, Leon, Mason, Naomi, Lucas, Lilia, Oscar, Iris C, Rosie, Stanley, Sicily, Olivia, Thibault and Séraphine.

Last, but never least, thanks to my gorgeous husband Reuben whose patience was tested, but never broken, and my wonderful children—Ava, Oscar and Archie who inspire me every day. You are the best.

(Not sure? Go to page 179 to work it out)